Millennial Spring

Designing the Future of Organizations

WITHDRAWN
LIFE UNIVERSITY
LIBRARY

LIFE UNIVERSITY LIBRARY
1269 BARCLAY CIRCLE
MARIETTA, GA 30060
770-426-2688

A volume in
LMX Leadership: The Series
George B. Graen, *Series Editor*

LMX LEADERSHIP: THE SERIES

Millennial Spring

Designing the Future
of Organizations

edited by

Miriam Grace
The Boeing Company

George B. Graen
University of Illinois (Ret.)

INFORMATION AGE PUBLISHING, INC.
Charlotte, NC • www.infoagepub.com

Library of Congress Cataloging-in-Publication Data

A CIP record for this book is available from the Library of Congress
http://www.loc.gov

ISBN: 978-1-62396-744-4 (Paperback)
      978-1-62396-745-1 (Hardcover)
      978-1-62396-746-8 (ebook)

Copyright © 2014 Information Age Publishing Inc.

All rights reserved. No part of this publication may be reproduced, stored in a retrieval system, or transmitted, in any form or by any means, electronic, mechanical, photocopying, microfilming, recording or otherwise, without written permission from the publisher.

Printed in the United States of America

We dedicate this career manual to our family millennials with love.

CONTENTS

PART I

PREFLIGHT

PART II

AIRWORTHINESS

PART III
EVENT HORIZONS

PART IV
NEW LANDS

PART V
A DECADE OF PROGRESS

PART VI

DEBRIEFING

PREFACE

This book reflects an interest in design technology and the collaborative design mindset by the two of us. Miriam Grace discovered the advantages of a design mindset in her systems architecture leadership for Boeing's Sales and Marketing functions. In parallel, George Graen discovered the advantages of conducting collaboration work employing a "design mindset," and asking "What if" questions about designing different work settings to produce innovations on many industries (Graen, 2002). We are proud to build on the cutting-edge, *Managing as Designing* by Boland and Collopy and report on the last ten years of significant progress in defining the design mindset. We are delighted to report that rather than a pessimistic practice of managers too often taking the safe mindset of choosing between the tired alternatives presented to them, today's managers are learning to design new alternatives. In this book, our experts in design mindset open the curtain to an even more promising view of our innovative future through design.

Although humans have limited rationality, we have found ourselves blessed with what we call *imagination* and *sentience*. These allow us to reach beyond our present to a future that may avoid much suffering and improve life. Our history is a story of great innovations that changed our self-made world. We have advanced to discover a primary design rule, namely, *if something exists in our world, it exists in some quantity and can be measured.* With this design rule, we are able to discover new ways to measure our imagined future and use design thinking to produce new prototypes to improve our life. Complementing this development, each new generation descends on their parent's world with a drive to replace the old and create a better world. Millennials are descending on our primitive workplaces and find them

Millennial Spring: Designing the Future of Organizations, pages xi–xii
Copyright © 2014 by Information Age Publishing
All rights of reproduction in any form reserved.

wanting. As a millennial, you were raised to expect a far different workplace, namely, a workplace similar to those you found at home, at school, and at college. Your so-called "helicopter" parent's generation redesigned your world for you, but have left it for you to do so in your workplaces. This set of essays was crafted with the hope that your parent's generation and yours might collaborate on the redesign of workplaces to make them more friendly and innovative.

In this book, we offer a way of thinking called "design science" that we find can be used to make workplaces more fun and innovative for all generations. In such workplaces, most everything is rationally negotiable. This includes those with parents, teachers, justice systems, social networks and cultural issues. In the new workplace, employees and customers are the most valuable assets of a company. All are entitled to life, liberty and struggle for happiness. Everyone deserves the benefits of the global internet and social networks tools. Finally, no one should "command and control" another at work, but should persuade based on reason and data. Does this sound too good? We sought out the best design thinkers of our day and invited them to a virtual seminar on designing both work and play organizations to be more engaging, productive and fun. We hope that reading our humble ideas will help spark those in power to move to design anew for one of the last areas of millennials disconnect, namely, the workplace.

We divided this book into separate sections to emphasize the scope of the design mindset. Many of the possibilities are being realized even without the proper words to describe them. For example, Michael Erickson struggled to describe his illustrative design process; however, after many abortive beginnings, he could not put what he did into polished prose. Instead of prose, he recorded his flow of consciousness. We take this as on encouraging sign that our design mindset can produce useful innovations through processes that we cannot yet describe. We hope that this book will be useful in extending your reach beyond your grasp.

We wish to acknowledge the editorial assistance of Joan Graen who kept our nonlinear process from complete chaos, our virtual team of authors for their genius, and special reviewing by Deborah Gibbons and Terri Scandura. We reserve a special appreciation to Michael Erickson for creating our innovative cover design. We also thank George Johnson, our publisher and friend, and his excellent staff for offering our work both as a complete set and as individual virtual chapters for high school and university courses. Finally, we dedicate this product to the extraordinary generations of the future.

—**Miriam and George**
Volume Editors

A NOTE FROM THE SERIES EDITOR

This edited monograph is the ninth volume in the *LMX Leadership Series*. All contributions were subject to rigorous review using APA research standards and revised accordingly. A decade ago, (2003) I conceived of this series as infecting all fields of human endeavor with the virus of my insights into organizational behavior and human performance by seeking out expert researchers in different areas of applied management research. I was convinced that I had found the "missing link" in managing a fully collaborative organization by constructing a network of business alliances including all people involved in getting the system to work. It is a capitalistic model of rewarding extraordinary network performance. The LMX "Pygmalion-like" alliances were found to have applications in many areas of management research. I began the series with a volume on the fundamentally necessary process of establishing and maintaining *equity* and *harmony*. From this base, I proceeded up the organization from two-person alliances to *units of alliance networks*, to *organizational alliance networks inside and outside*. The last two volumes were focused on the special conditions of first responders in harm's way and on methods to manage complex system changes and making decisions in novel situations. This present volume is *predicting the future* using collaborative design mindsets. I hope you have enjoyed the journey as much as I have. This series has been a labor of love for me and my wonderful partner, my wife of 55 years, Joan Graen. We are indebted to our extraordinary publisher, George Johnson of Information Age Publishing, and his staff who have been our alliance partners for over a decade.

Cheers!

—**George Graen**

Millennial Spring: Designing the Future of Organizations, page xiii
Copyright © 2014 by Information Age Publishing
All rights of reproduction in any form reserved.

FOREWORD

Call for an Under-30s Revolution

The world is changing fast, and the *millennials*—the generation of people who became adults around 2000, or in the decade or so after—have been right in the middle of it. From Independence Square in Kyiv to the streets of Caracas, from Taksim Square in Istanbul to Zuccoti Park in New York, and from Silicon Valley to Wall Street, it's the 30-and-under crowd courageously leading the quest for different ways. Less invested in past approaches, tech-savvy to a fault, and painfully aware of the challenges left to them by earlier generations, they're not willing to "settle"—to make the same compromises (and mistakes) they think their parents made. And although they sometimes get rapped for being self-centered, all the evidence I see—and I've taught thousands of them on two continents, and even have one in my own family—suggests that the millennials represent real hope for the future.

This book proposes *design* as the *means* to a better future. To remain relevant when faced with the new realities of continual change, people must change too, and also change the world around them. As suggested by Boland and Collopy in their 2004 book, *Managing as Designing*, the diverse collection of ideas and approaches clustered under the heading *design* constitute an increasingly effective path to favorable change. But this book goes further: It suggests that collaborative, humanistic design, deployed in communities of peers learning from each other (and from the mistakes of recent history), might just position millennials to become better stewards of Planet Earth than their baby boomer parents have been.

Millennial Spring: Designing the Future of Organizations, pages xv–xix
Copyright © 2014 by Information Age Publishing
All rights of reproduction in any form reserved.

xv

Design is not new, of course; it has long been involved in the making of new "things" in the world: tools, frameworks, systems. The 20th century saw the making of some marvelous things. Tools, drugs, agricultural systems, architectural structures, ways of organizing people and work. Many of these "made things" were undeniably successful in terms of their potency—their ability to change the world. Systems of mass production, to choose just one example, have created a standard of living that the world has never seen before (although the benefits are not evenly distributed). And whether it was acknowledged or not at the time, design played a primary role in all these marvels. As Dave Kelley, founder of the celebrated firm IDEO, has observed, the only thing in our surroundings that wasn't designed by somebody is nature. Because of design, lives got saved, lives got better, some people even got rich.

But the 20th century also brought potency in harmful forms. We made weapons powerful enough to destroy the planet. Many of the systems we made had big, unforeseen side effects, outcomes that ran counter to what we would have chosen: climate change, financial disruption, species extinction, and new diseases, some of them behavioral and social. Some things we made in the 20th century are so potent that they now threaten us.

This inheritance has not gone unnoticed by the millennial generation. Everywhere you look, there is growing sense, especially among young people, that to address the grand challenges of this new century, we need different ways. Environmental catastrophes (Deep Water Horizon), nuclear disasters (Fukushima Daichi), a world brought to the brink of financial ruin (Wall Street, September 2008) . . . these are outcomes of the technocratic approaches of the *last* century, taken to their expert extremes. It is the old ways that have yielded the current challenges. Clearly, then, we need to do something different. We still need design to save and improve lives, maybe even make some people rich, but now we also need more from it than that. We now need it to help save the world.

In the chapters that follow, a most capable set of authors, building on deep insights from past work, especially that captured in Boland's and Collopy's *Managing as Designing,* describe productive directions for designing the future. I'll leave it to these authors to describe the details of their new ways. But some common threads seem to me increasingly visible, borne of the inspiration provided by Boland and Collopy, and they offer hope that we might successfully address our grand challenges.

Boland and Collopy proposed, as a framing concept for their 2004 book, that we adopt a *design attitude* as we address future challenges. They contrast design attitude with the *decision attitude* that has dominated management thinking in the past several decades. They trace the decision attitude to Herbert A. Simon's 1969 book, *The Sciences of the Artificial,* arguably the first treatment of design as a way of thinking.

I had the privilege of studying at Carnegie Mellon with Herbert Simon, and there can be no doubt of his immense contributions, to both design and management thinking. But it's also undeniable that his research, and much that came after it, included a framing based on decision making and problem solving. Simon's famous course at Carnegie Mellon was called "Human Problem Solving" (HPS); that was also the name of his influential book with Allen Newell. Simon's earlier work, such as his book *Administrative Behavior*, which vaulted him to prominence, and his later classic with James March, *Organizations*, both focused on human limitations in making decisions. Partly due to Simon's influence, framing management challenges in terms of "decisions" or "problems" has become widespread, and it has been useful in many contexts.

But not everywhere. Framing conveys with it hidden assumptions. Some acts of design can be conceived in terms of problem solving, and design is, of course, often motivated by concern about particular problems. But there has always been a question of whether *making*, especially in its most creative forms, can be adequately described solely in terms of decisions or problem solving.

As Boland and Collopy observe, "a decision attitude carries with it a default representation of the problem being faced, whereas a design attitude begins by questioning the way the problem is represented" (2004, p. 9). That's a vital insight: We need to remake *problems too*, if we want to arrive at truly new ways of interacting with the world. Rejecting or remaking the problem, as formulated, is a powerful part of the act of creation. Genuine creativity, especially in less domesticated variations, has a potency that can, if we have the right *attitude*, propel us past the tidy boundaries suggested by a predefined problem formulation.

When we adopt a design attitude, we get to new problems and solutions by *doing*, often trying one thing and then another, in a process that Larry Loftis, a Boeing executive, once described to me as "trystorming." Trystorming, a relative of brainstorming, requires not just coming up with ideas, but actually *doing* them, cheaply and rapidly. In the course of such a process, unexpected things happen, some of them quite valuable. New problems and their solutions *emerge*. Boland and Collopy quote architect Frank Gehry, "If I knew how a project was going to turn out, I wouldn't do it." His implication is clear: Outcomes arrived at through a more intentional, less emergent process, would be less valuable.

Technology plays an increasingly important role. Trystorming requires that trying is not prohibitively expensive or time consuming; if it costs too much or takes too long to try things, you won't be able to afford to try very many things. And, as Boland and Collopy lament, a lack of money to explore further alternatives has been used to excuse a lot of mediocrity in the world: "[T]he decision attitude toward problem solving that dominates management education, practice, and research favors default alternatives and locks us into a self-perpetuating cycle of mediocrity."

But if technology can be applied—in simulations, automated testing, or in support of prototyping, for example—to reduce what Lee Devin and I have called "the cost of iteration," then you *will* be able to try many more things. Valuable outcomes will emerge, in part, simply because you will feel free to try out many more ideas when you can cheaply and rapidly iterate. But there's more too it than that. When you can try many things cheaply and rapidly, you'll also stumble onto more realizations, by accident. Not only will you try more things you *can* think of to try; you can also stumble productively onto valuable possibilities that you *can't* or *didn't* think to try. Technology both enables and provides new venues for this kind of productive stumbling; Boland, Lyytinen, and Yoo, describe, in a 2007 paper, how digital platforms provide "zones" of collaborator encounter, wherein actions combine to "create unpredictable patterns of interactions when they collide with one another."

Designers and architects have long used low-tech approaches (sketching, modeling with cardboard, styrofoam, plywood) to help them work with a design attitude, so such an approach to making doesn't arise purely from the application of digital technologies. However, proliferation and increasing application of these technologies makes the design attitude increasingly relevant, more often the best approach in many more situations. Also, communication technologies, most notably the Internet but also mobile phone networks, provide rapidly evolving and proliferating social platforms, that can be used to coordinate, interact, and collaborate. Tech-savvy millennials are obviously ideally situated to run with technology-assisted approaches.

Indeed, the evolving ideas of collaboration and community that have been embraced by millennials are perhaps most important and exciting of all reasons we might hope to improve upon past ways. Increasingly, activity aimed at improving the world is *participative* and *democratic*, involving within making processes the people who have a stake in their outcomes. Innovation has become *customer driven* and *user orie*nted; often it involves *crowds*, whether physical or virtual. The web serves as the vital instrument that facilitates such productive activity and the venue for the community that engages in it. Distance offers less impediment to the formation of groups and communities than ever before in history. Ideas travel fast, and change is contagious, which prompts dismay and fear from those who cling to old ways.

Yes, the world is changing rapidly, and everywhere you look millennials are the drivers. As I write, a community called "Maidan," which counts many under-30s among its numbers, continues to occupy Independence Square in Kyiv, while the opponents of change take the southern most parts of that country by force, with tanks and helicopters. Timothy Snyder observes in the *New York Review of Books* that, although the word *maidan* is old, with origins in the Arabic word for "gathering place," the concept has evolved rapidly. Maidan is "not just a marketplace where people happen to meet, but a place where they deliberately meet, precisely in order to deliberate, to speak, and to create a political society." Interestingly, the word has not, historically, existed in the language of the ruling class that opposes reforms in Ukraine.

But it's an idea that has now entered all languages. Call it what you will, a change-oriented community, using better approaches, doing work facilitated by technology, especially the Internet, is joining virtual hands around the world. The forces of the past still marshall opposition, often with surprising effectiveness. But I know where I'm placing my bets.

So consider this a call to all millennials—here, in the pages of this book, are some of the means. Get out there and create the under-30s revolution. Solve the problems your parents couldn't. Do it together, with a conscientious eye to what works for all involved. Get out there and save the world.

—**Robert D. Austin**
Professor, Management of Creativity and Innovation
Copenhagen Business School
Co-author of *The Soul of Design* and *Artful Making*

REFERENCES

Austin, R. D., & Devin, L. (2003). *Artful making: What managers need to know about how artists work*, Upper Saddle River, NJ: Prentice Hall.

Boland, B. J., & Collopy, F. (2004). *Managing as Designing*, Stanford University Press.

Boland, R. J., Jr., Lyytinen, K., & Yoo, Y. (2007). Wakes of innovation in project networks: The case of digital 3-D representations in architecture, engineering, and construction. *Organ. Sci. 18*(4), 641.

Newell, A., & Simon, H. A. (1972). *Human problem solving*, New York, NY: Prentice Hall.

Simon, H. A. (1969). *The sciences of the artificial*. Cambridge, MA: MIT Press.

Snyder, T. (2014, March 20). Fascism, Russia, and Ukraine. *New York Review of Books*. http://www.nybooks.com/articles/archives/2014/mar/20/fascism-russia-and-ukraine/.

PART I

PREFLIGHT

CHAPTER 1

A BRIEFING[1]

Miriam Grace
Boeing Company

George Graen
University of Illinois (Ret.)

This book is designed to bring you along on a flight into alternate futures. Please keep your seat harness firmly locked and your disappearing apps stored during take offs, landings and weightless periods. Our mission is to open your minds and hearts to the wonderful tools for improving our quality of life from educating perspective parents and their children in child rearing through college, career, retirement and the completion of life with dignity. We can make available to everyone better ways to do most everything. We know that most of our tools and ways to use them were the product of past trial and error. They were not designed (typically) by having empathy for the problem in depth and/or considering multiple possible alternatives. Man, by nature, is pragmatic and seldom initially designs the right tools for the job. This is changing.

In an age when we have the tools to mine the secrets of the universe and see back to the beginning of time (Hubble Telescope), we are learning to leverage tools to explore what is an even more mysterious cosmos, our human capacities. Exploring these capacities and figuring out how to employ them in service of individual and organizational goals is the purview of leadership and, we would argue, the tool leaders can depend on to do this

Millennial Spring: Designing the Future of Organizations, pages 3–11
Copyright © 2014 by Information Age Publishing
All rights of reproduction in any form reserved.

work is design. If you are willing to explore the notion that modern corporations are the most powerful institutions governing modern life, then what affects these institutions affects the whole. Designing organizational systems for the future, therefore, becomes vitally important work both for those working inside those systems and for the stakeholders.

In this book, we have asked a virtual team of organizational-designers, -engineers, -information technologist, -managers and -psychologists to look into the future and predict the impact design will have on organizational systems over the coming decade. A decade back, Richard Boland and Fred Collopy (2004) glimpsed the future employing something that they called the "design attitude." They appropriately proceeded to acknowledge the pioneering work of Herbert Simon, (1976) Nobel Laureate in economics and thought leader in matters of the talents and limitations of both humans and computing devices. Simon's work comparing human and artificial intelligence of computers enabled him to discover that the design of searching for the needle in the haystack made a difference employing either human or computer thinking. He had the breakthrough insight that the needle was the designerly identification of the root problems. As Boland and Collopy point out "a decision attitude carries with it a default representation of the problem being faced, wherever a design attitude begins by questioning the way the problem is represented" (2004, p. 9). This insight proved meaningful for innovation.

This book, appearing a decade later, adds to the themes explored by our forefathers in 2004 with the benefit of hindsight and new tools. For example, when environmental change renders a faithful system obsolete, a decision must be made by management either to band-aid the old or innovate a new system. But, how can a new system be designed when a number of different conditions have changed? Which are the root causes of the present system failures? Fortunately, you can learn a design discipline that helps to avoid the default of needing to proceed by trial and error and repeatedly solving the wrong problem. This default may eventually succeed by a process of elimination, but at the expense of time, resources and overlooking the diamonds.

Systematic improvement of systems by creating new ones when old ones become obsolete using the *design attitude* is becoming accepted by business. We have developed new protocols for the discovery phase of research relations (Meehl, 1977). We have been searching for such systematic protocols for the discovery phase since we realized that careful observation of the secrets of the man-made world began to yield ideas that were consistently validated. This book suggests the breath and depth of the applications of the design attitude. An example of a minimalist description of this attitude might be systematically researching the question: What if we built a structure that would permit instant access to the world's knowledge by households? Those employing the design attitudes begin by investigating all aspects of this question by describing the actors, their behavior and their context—the ABCs of specification. As we now know this process of development produced Wikipedia. Natural science's strength is demonstrated in the search for man's understanding of our world, where the design attitude strength is aimed at improving the operation of man-made adaptations to our world. Design questions become relevant when established systems become dysfunctional or obsolete due to changes in conditions. It is the search protocol for innovation in security, social service, health, education, psychology, economic, and world peace to name a few. Recent candidates for the design attitude would include health care, immigration, government collaboration, education the design of work systems friendly to new generations of employees and the very careers of those who design. Too often our top leaders in the above arenas are certain they know what kind of innovative system they want only to find out latter that their new system does not work and it's back to the drawing board. Proper design thinking as described above is designed to make the new system successful the first time, because it is based on proper investigation of its proposed context. For example when an artery bridge in Minneapolis on August 1, 2007 fell into the Mississippi carrying traffic, we witnessed a failure in design attitude. One size seldom fits all conditions.

DESIGN THEMES

Recall that the themes explored by Boland and Collopy's set of thought pieces in their *Managing as Designing* (2004) were:

- Managers must solve problems and act on time and budget.
- Design attitude is at the core of strategizing development, and collaborative cascading of execution.
- Managing design is a collaborative process.
- Unmet challenges for the next decade on a global scale for better organizational environments.
- We are often trapped by our vocabulary.
- Using multiple models of design problems is helpful to stay fluid.
- Sketching, mapping and storytelling also aid staying fluid.
- Beware of pet ideas.
- Seek widely functional designs.
- Break from default! (adapted from pages 17–18)

As one major gift derived from *Managing as Designing*, to our book was that we could compare progress to the past decade of contributions to the art and science of design. Clearly, our authors stand on the shoulders of creative and purposeful talents. We have discovered that when "artistry" and "functional" describe a product or service people flock to possess it. We find that our understanding of this new approach has advanced significantly in the last decade. This marriage of the mysterious world of the artist and the wonderful world of science presents different means of understanding ourselves at work. Graen was trained in logical positive science and practiced the same throughout his career until he was introduced to the approach of design. Since then, he has incorporated the power of design thinking into his Leader-Member Exchange (LMX) collaborative research and theory development (Graen, 2013a, b). This book reveals the potential of whiz kids who will achieve "star" status from this marriage of art and math that we discuss as design. All of us have our inner artist struggling to be recognized. This idea is not new, but our improving understanding seems innovative. You, as a young or even a mid-career professional, can fast-track your career with these "collaborative-design" competences.

AIRWORTHINESS

This book proceeds with Miriam Grace, Executive Information Architect at Boeing, looking into the future of organizational systems within a business context. She lays out the progress made in the last decade to evolve one of the few organizational competencies that is proven to directly drive innovative behavior. She provides a clear roadmap for how you can enhance your professional career opportunities by focusing on a strategic set of capabilities and framing them in a context of organizational systems renewal. As a master craftswoman of the art and science of creating and implementing innovative designs, she has seen both the triumphs and tears from designing new and improved man-made systems. She finds that the best designed products are the work products of networked collaborative design teams, who continuously learn from each other during the iterative process of designing. Miriam recommends practical methods, tools, and mental models you will need to contribute effectively inside an innovative design team culture. Ultimately, she challenges you to lead from a whole system design perspective and enable virtuous feedback loops in your organization to foster and sustain a holistic culture of innovation and creativity.

George Graen, Center for Advanced Study at The University of Illinois, C-U (ret.), suggests that new designs are being constructed for entire organizations to overcome inefficiencies and the dehumanization aspects of hierarchical power. He recommends that you realize the global value of *collaboration* through designed alliances in both business and voluntary enterprises. His leadership-motivated excellence processes of collaboration among people and across organizational boundaries allows people the respect and trust that underlies lasting interpersonal growth (Graen & Schiemann, 2013). This breakthrough approach Graen calls the "collaborative-design mindset," promises improvements in global understanding through active team participation in purposive collaboration.

Andrea Cifor and Sarah Mocke from Microsoft discuss realities you are undoubtedly facing if you are in an organization that is feeling the whiplash of technological change—and who is immune from that in this new millennium? This chapter gives you an insider's insight into the commitment to change and disruption that comes with a decision to pursue a technology-focused career direction. But, you also experience the angst of those who have achieved expert status in their fields and who still find themselves at-risk and vulnerable with ageing networks and shrinking options. The ability to be a versatilist, to diverge and consider situations from multiple perspectives, to develop deep empathy for the customer, to collaborate in all things may be counter-culture to your current organizational culture, but the authors identify them as competencies that can break a fall from expert status or accelerate your trip to that status—these are design competencies.

EVENT HORIZONS

Min Basadur, a business consultant and professor at DeGroote University, gives us the practical application of his theory, bringing concepts to life through real stories of engagement. He details the process of inventing and learning that provides a roadmap to organizational survival in the new millennium. His synthesis of paradigm-shifting, learning and designing change through a structured innovation process beginning with problem finding, sensing, defining (prior to problem solving) and ending in implementing change makes this new technology digestible and very doable.

Colleen Ponto and Peter Coughlan from the Bainbridge Graduate Institute have designed a breakthrough composition of design and systems thinking, which is profound in its simplicity. They clearly make the case for the synthesis of design and systems thinking in order to address the human interaction dynamics and emergent problem spaces that populate today's complex work environments. Ponto and Coughlin chart a path for effective leadership in this new world as a design-in-action model where leaders operate to develop and facilitate a constantly in-work design to achieve organizational direction. The example they provide of how they are evolving this method designed to shift organizational forms is easy to follow and replicate and provides techniques that you can immediately apply in your work.

Jim Hazy and Tomas Backstrom describe their collaborative leadership design that they developed from their work at Mälardalen University in

Sweden. Their chapter expands the conversation by bringing forward the complexity perspective. They posit that the reality of malleable and ever-changing processes, structures, and technologies that drives the need to continually reshape perceived problems requires a new kind of leader for this new millennium.

NEW LANDS

Deborah Gibbons from the Naval Postgraduate University promises to un-cover your cultural predispositions about managing across cultures and to educate you on how to influence your organization to make a positive impact on its international employees, partners and communities. This chapter is a deep dive into a cultural competencies knowledge base that composes the essential elements of this multidimensional subject, and it's an easy read.

Marcus Jahnke and Ulla Johansson Skoldberg from the University of Gothenburg enlarge the context by discussing the creation of meaning in design praxis. In a recent field experiment funded by the Swedish Gov-ernmental Innovation Agency, they studied "non-designerly" companies from a variety of industries in the act of integrating design as an approach to innovation into their daily operations. In these experiments artistically trained designers involved multi-disciplinary groups in experiencing design hands-on. The question of what design as a practice brings the rationale for engaging designers in the processes of business, and the far-reaching chal-lenges for managers are discussed in relation to stories from the different cases. Their research is one of the few recent explorations of the value-add of design within operations processes and in varied business contexts.

Ben Zweibelson, (MAJ. US Army), Grant Martin, (LTC. US Army), and Chris Paparone, (Col. US Army), officers in the U.S. Army, share how de-sign thinking is still very much a "toe in the water" of Army operational art, but also describe how it is misinterpreted as design seeps into Army manuals. These authors describe their efforts to heighten understanding of design thinking and how well the design mindset better suits the patterns of disruption that characterize modern warfare. Their stories, some of which come fresh from the battle fields of Afghanistan and Iraq, tell of teach-ers who are serving as "early adopters" of design science. They share their journey to enable the Army to see and exploit the competencies of design methods and the adaptability of a design mindset—a survival skill-set for the next generation of military practitioners.

Skip Rowland, a whole systems designer who consults with entrepre-neurs and multicultural small business owners in the urban core, took on a project intended to help military service members transition out of the military and successfully enter civilian life and the civilian workforce. Skip shares his personal story and some of the stories of the soldiers he works with who are essentially mid-career professionals They may be like you, who have a very different work experience from most people in the job market, and who are getting that sinking feeling that finding a living wage job may be the toughest fight they've ever been in. Skip helps veterans understand and adopt an "entrepreneurial mindset" to meet the challenges of shifting cultures. He works through existing organizational structures in military and civilian organizations to create a bridge of shared responsibility between military and civilian cultures and values, which in turn, creates a bridge to jobs for veterans.

Collopy and Boland revisit their ideas of a decade ago and conclude that design has been strong, but not strong enough. They suggest that we probe deeper into our human-made world and develop and employ our creativity to improve our future. They advocate for "strong design" that emphasizes a synthesis of design action with design thinking for an integration of left brain and right brain functions into holistic solutions. Their message for the millennial manager is to develop a strong design attitude, with strong design skills and a strong sense of design space. With those three qualities, the new millennial generation can indeed do better than we have done before.

DEBRIEFING

Michael Erickson is saved for last. He created our cover design. Michael is from Boeing. He describes himself as a "systems analyst that draws" and he will inspire you to improve your visual communications skills. The relaxed style of this chapter and the graphics that punctuate the pages fly cover for the author who has flipped the script on himself, recording his ideas and reflections in text and revealing how his thoughts run all the time he is creating visual representations of others. His direct style will engage you and his graphics are fun. He takes you on a flow of consciousness of a design-artist. When you're done, he will have taught you how to recognize and surrender to the emergent dynamics in life. Design doesn't just happen! "Come ride with him into the other side of the creative element, to a place that may be a little less academic, and certainly more experiential, where we may see if we too can catch a glimpse of the potential we all have to leap beyond our egos, certainly beyond our fear of ridicule, into dimensions where, if we dare, join with those unruly elements of our existence (that terrifying aspect we call 'the chaos') where we begin to co-create in it the answers that are beyond our expectations.

DISCLAIMER

Projecting the nonlinear future is a risky venture for a new approach from art and architecture such as collaborative design. Designers must get it right the first time for their patrons and stakeholders or suffer historic abuse from present and future generations. When a tower leans too much or falls or becomes a white elephant, the designers reputations also crash. Designers have learned by their mistakes and success that the design must be right the first time. Unfortunately, those who design special purpose work environments for people typically copy what someone tells them is the "best practice" and design one to fit all. Sponsor's gut feelings often establish the parameters. The good news is that in the last 10 years, design thinking has emerged to question "best practices" as universals. Designers realize that what may work in a particular environment may not work even in similar environments. Moreover, today's designers emphasize who the actors will be, what they will be doing, and where they will be doing it. To accomplish this, designers must empathize with these actors, the behaviors in their home organizations and ask, what conditions would enable the workers to make their space more friendly and productive?

WHY MILLENNIALS RESIGN

"It will be necessary to transform the core dynamics of the workplace". (PWC, 2013). The time has arrived to design large firms friendly to the colleagues who were treated as gifted from the cradle through professional education. Alerted by alarming trends in resignations of new hires at two years, Price-Waterhouse-Cooper, LTD. contracted with the University of Southern California and the London Business School to perform the largest survey of a private firm. They found that there has been a sea change which requires a new design for the millennials entering the work force.

- Design of careers was unattractive
- Concept of work should be a "thing" not a "place"
- Workplace culture should create "teamwork and community."
- Most stereotypes are dysfunctional (adapted from PWC, 2013, pp. 8–9)

Gratifyingly, the recommendations of this comprehensive investigation generally are supportive of those described in this book.

CONCLUSION

Today, large corporations are buying entire design firms (Hurst, 2012). They think that their customers and their mother organizations require full range and continuous design service. Rather than form a business alliance with a design firm, corporate executives find that design thinking is the future. They are placing design people on their top management teams (TMT). Moreover, this trend is international and local. Even the decision systems content of MBA programs is experiencing pressure to enhance the content and new tools of design thinking (Grace & Graen, 2014). This book introduces new perspectives of understanding the world as if it were a new ride at Disney for patrons that are not seeking simple thrill rides, but are composed of readers like you... readers that are ready to buckle in for a new experience to take your minds to different slipstreams of understanding reality. This is the essence of game-changing design!

NOTE

1. Our virtual team approved this briefing. Editors' contributions were alphabetical and equal.

REFERENCES

Boland, B. J., & Collopy, F. (2004). *Managing as designing*, Stanford University Press.

Grace, M., & Graen, G. B. (2014). What if we designed A MBA for the Future? *Decision Science*.

Graen, G. B. (2013a). Overview of future research directions for team leadership. In M. G. Rumsey (Ed.), *The Oxford handbook of leadership* (pp. 167–183). Oxford, UK: Oxford University Press.

Graen, G. B. (2013b). The missing link in network dynamics. *The Oxford handbook of leadership* (pp. 359–375. Michael G. Rumsey (Ed.). London, UK: Oxford University Press.

Graen, G. B., & Schiemann, W. (2013). Leadership-motivated excellence theory: An extension of LMX. *Journal of Managerial Psychology, 28*(5), 452–469.

Hurst, N. (2012, May). Large corporations are buying design firms. *Industry Week.*

PWC. (2013). PWC's next Gen: A global generational study. www.pwc.com.

Meehl, P. E. (1977). Specific etiology and other forms of strong influence: Some quantitative meanings. *Journal of Medicine and Philosophy, 2*, 33–53.

Simon, H. A. (1976). *Administrative behavior: A study of decision-making processes in administrative organization* (3rd ed.). New York: The Free Press.

PART II

AIRWORTHINESS

CHAPTER 2

WELCOME TO THE FUTURE OF ORGANIZATIONS

Design, Analytics, Innovation!

Miriam Grace
The Boeing Company

ABSTRACT

Your professional future will be significantly influenced by how well you position yourself and your organization for achieving and sustaining competitive advantage. This chapter looks forward and predicts the convergence of several emergent business strategies that will profoundly impact your career and the organizational systems you work within over the coming decade. Highlighted is a framework that unites design thinking and design methods with advanced analytics as significant influencers of business innovation. These three business capacities (design, analytics, and innovation) are already well recognized individually as viable global business differentiators. What is new here is the potential for breakthrough advantage by integrating these separate capabilities into a holistic strategy for creating an innovation incubator inside your organization. The framework is offered as a template to represent the synergistic interrelationships. The framework described also highlights the impact of these activities on an organizational culture when the triad is

Millennial Spring: Designing the Future of Organizations, pages 15–40
Copyright © 2014 by Information Age Publishing
All rights of reproduction in any form reserved.

implemented with a whole system design intention. The resultant virtuous feedback loop, created by an organizational culture that fosters innovation, will reinforce and sustain the whole system.

INTRODUCTION

You're a millennial, early in your career, or a mid-career professional feeling the foundations of your career shifting beneath you. You're looking for that edge, that knowledge, skill, or capability that will make you a standout in your field or, perhaps, allow you to be more of a versatilist—one who can take on multiple roles—a definite professional advantage in this uncertain and constantly changing job market. You need to be able to apply this new knowledge or skill in a variety of contexts, so it needs to be foundational knowledge and new learning skills, not specific to any one profession, technology, or industry. It needs to be new, fresh, and relevant so that it sets you apart and gives you unique value or puts you a rung up in your search for career opportunities. You need, in short, to hear the signal amidst the noise (Silver, 2012).

This chapter discusses three emergent phenomena that rise above the noise of business-based media coverage like dominant signals in a word cloud: Design! Analytics! Innovation! They stand out, and this chapter explains why their prominence in the media isn't hype. They are capturing attention because each, individually, is bringing new and significant value to enterprises in novel ways. However, their implementation in daily business practice is typically fragmented and thus their synergies are lost advantage. What you should be asking is "What if they were linked together intentionally, in an integrated and holistic way?" Consider how you could leverage such insight. Let's see: (1) Capabilities in any one of these areas of expertise could be applied to your advantage in any of the main institutions of life (political, social, educational, and economic)—*Check!* (2) Knowledge about these capabilities and their application in a business context could deliver value across a wide variety of industries and professions—*Check!* (3) Any one of these, if learned, adopted, and perfected through professional practice, would create competitive advantage—*Check!* (4) Understanding how they work together, however, is where the big payoff results for your career, for your organization and fellow employees, and ultimately for the future of organizations as a whole—*Double check!!*

The discipline of design, which instructs on the purposeful creation of things, and the mind-set of design thinking, both a paradigm and a rich set of methods, have proven themselves to be competitive differentiators over the last several decades as design has moved up the business value chain from enhancing the aesthetic appeal and style of products, to taking center stage in new-product development, and within the last decade, to proving its worth to corporate innovation initiatives. Analytics, "the extensive use of data, statistical and quantitative analysis, explanatory and predictive models to drive decisions and actions" (Davenport & Harris, 2007, p. 12) has been around for decades under the reporting and business intelligence rubric. What is emerging over this decade is a different animal that has the potential to transform business-technology strategies across the world. Innovation, defined as "people creating value through the implementation of new ideas," (Innovation Network, n.d.) has taken the center of the global business stage as enterprises wring the last big wins from the quality and productivity movements. Establishing a culture of innovation inside an organization and casting that cultural network externally to join supply chain partners, universities, social networks, and beyond—and knowing how to mine that network to advantage—create one of the richest renewable resources available to businesses today. A business that is intelligent about exploiting its intellectual property, like information, is also an attractor for just the kind of talent that will become increasingly difficult to secure as organizations become more focused on data acquisition and exploitation. The triad of design, analytics, and innovation form a virtuous feedback loop for continuous organization systems renewal that millennial managers can promote to attract and retain "stars" in their talent portfolio; but understanding that systemic capacity requires some context setting.

The next three sections establish that context by describing each of the individual components and how they can work together in a deliberately integrated way. Simply leveraging them as a set of integrated methods and tools will positively impact your organizational culture, as illustrated by the framework offered in Figure 2.1. This framework, supported by the foundational knowledge offered in this chapter, is a strategy set for a game-changing millennial manager, like you. Understanding the synergies among these three business activities (design, analytics, and innovation) and how their integration might impact your organization gives you an edge. Having a package of integrated methods and tools in a critical area of business like data analytics is going to drive some level of innovation because the integration tends to break down silos which means people spend less time searching and patching and more time gaining insight. That kind of flow through the innovation pipeline is going to have a positive impact on your organizational culture.

If you choose to have a hand in guiding organizational change in your organization, you should at a minimum be educated in the principles and practices of each element of the framework and how they can interoperate and support each other as a part of a breakthrough business strategy. Perfecting your ability in one or more of these capabilities will differentiate you from your peers. Facilitating their integration will get you recognized as a change leader.

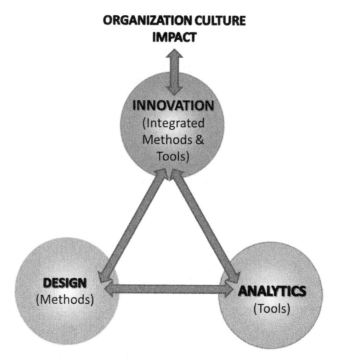

Figure 2.1   Integrated design, analytics, and innovation in practice. © 2014 Miriam Grace.

The three sections that follow briefly introduce the capabilities, set the context for how they interoperate within a work setting, and suggest ways for you to engage to your advantage. But the big payoff is described in the final section of this chapter. There you will find an evolution of the framework that describes a holistic perspective on the integrated power of design, analytics, and innovation; a power you can intentionally leverage to drive the culture change your organization needs to be competitive in the new millennium. But first, let's look at each element in the framework so you will understand their individual contributions in terms of business value delivery and the synergies of implementing them as a unified strategy.

DESIGN

Context

The discipline of design consists of a well defined set of principles, processes, and methods that have been in constant use since the days of the

Bauhaus Design School of the early twentieth century (Lekach, 2013). The emergence of design as new curricula in business schools and the employment of the discipline of design outside of product development R&D functional areas within businesses have been steadily gaining attention as mitigating responses to the challenges of the global marketplace.

In 1997, Bolman and Deal described the properties of organizations as "complex, surprising, deceptive, and ambiguous" (p. 22), and that description has only become more characteristic of our twenty-first century complex and interconnected lives and work spaces. Over the last two decades of extraordinary change and increasing business complexity, however, the need for increased employee engagement and collaboration, and initiatives to increase productivity and effectiveness, began to drive decision making to the lowest levels of the organization, creating a shift in the locus of management control and a resultant change in the level of knowledge investment in each individual employee. A tipping point was reached when the business focus shifted from employee empowerment to innovation. Organizations began looking for ways to mine ideas from their employees, and that shifted executive and management attention from the outputs of work to the front end of processes—to the information-gathering and discovery phase of solution finding—an area of work processes that has been underexplored and therefore underexploited.

Understanding design as an innovation accelerator has exploded the scope of recent design research. For example,

> The United States has launched a national design policy initiative to monitor and understand the role of design in the national and global economy, and the European Union is overseeing a series of public consultations on how the EU can further support design-led innovation with the aim of integrating design into innovation policies. In the UK, the Design Council has performed a series of studies on design's role as a strategic instrument to maximize performance and trigger innovative processes, even during periods of crisis. (Celaschi, Celi, & Garcia, 2012, p. 6)

Design has shown a capability for taking creativity beyond products and services and into the creation of new "behavior spaces" for enhanced experience in the workplace for employees and enhanced customer experience of designed products and services. The idea of managers as designers (Boland & Collopy, 2004) introduced the possibility that a true cultural revolution may be possible that would fundamentally change how we intentionally organize (or disorganize) ourselves in organic and self-sustaining ways in the future while still accelerating and sustaining a healthy value proposition for business.

Value and Purpose

Whether we are talking about computer systems development, seeking productivity improvements, reducing waste by applying Lean methods across a value stream, or other everyday improvement operations in our organizations, it is the front end of the process where we are often weakest, and not surprisingly where we have traditionally put the least effort. But experience has shown that this is the area where we should seek deep insight about the context of a problem to ensure we are working the right problem and developing a shared understanding of the opportunity elements. Design methods guide us to observe the interconnected elements in a problem space to identify what generates the problem, and recognize and explore the patterns of activities, interrelationships, assumptions, beliefs, constraints, and other elements that are working to hold a problem situation in place. Design is particularly effective in situations where there are multiple dimensions to consider, layers of meaning that challenge linear logic, and many gaps in understanding. Design offers many methods, aids, and tools to tackle the level of complexity inherent in this vitally important part of any solution-finding process. What this means is that design provides a comprehensive guide to one of the thorniest problems of the innovation movement; that is, *how to do innovation.*

Design has a multidisciplinary set of principles and methods for "gaining insight about people and their needs, building strategic foresight, discovering new opportunities, generating creative possibilities, inventing novel solutions of value, and delivering these into the world" (Wasserman, 2013, p.1). Design is pragmatic in that it develops knowledge that serves purposeful action. It is inclusive, collaborative, and multi-perspective, and teaches the art of participative discourse and iterative experimentation. Design drives a human-centered, empathetic, and deeply collaborative innovation process that synthesizes and balances the disciplines of science and the humanities, requiring the synthesis of professional capabilities such as technical and business architecture, engineering, social science, communications, strategy, finance, and economics.

Roger Martin (2004) refers to "the design-thinking organization" and how such an organization "applies the designer's most crucial tool to the problems of business [and] that tool is abductive thinking" (p. 62). Martin clearly differentiates deductive logic, "the logic of what must be [that] reasons from the general to the specific . . . and inductive logic, the logic of what is operative [that] reasons from the specific to the general" (p. 63), from abductive logic, a concept originally brought forward by the American philosopher Charles Sanders Peirce (1839–1914). Peirce posited that new ideas come into being via logical leaps of the mind where a person notices something, wonders about it, and through the processes of the brain continues to work in this train of thought, at times consciously but also unconsciously. Often this process results in an "ah hah!" experience that seems to come out of nowhere. Martin explains that this process is "not declarative reasoning" (p. 65) as we are used to in our scientific-based analytical reasoning world. He emphasizes that the goal of abductive logic "is not to declare a conclusion to be true or false. It is modal reasoning, its goal is to posit what could possibly be true" (p. 65).

One can appreciate that this kind of logic has great appeal and can be understood to have business value in settings like research and development labs where "star" inventors are encouraged to make the mental leaps that come from continual exercising of abductive thinking processes. However, imagining this kind of thinking spreading broadly across organizational cultures has kept it in the shadows. Traditional organizational structures and historical management learning paradigms, and their resultant attention to control of resources and quick delivery of solutions, may have driven twentieth century managers to see this as a potential risk scenario.

But today, imagination and initiative are understood to be the critical drivers of business growth in this new, interconnected world. Business is on the hunt for ways to continually exceed customer expectations or, even better, to anticipate what their customers need and even address customer sets that are in adjacent businesses or totally unrelated product or service contexts. What makes design so appropriate for this business scenario is that it is a discipline that is itself configured to address emergent phenomena—like customer needs.

From this perspective, it is easy to see how well design fits into a millennial management paradigm. Design covers the required mind-set and methods, but there must be a focus for operationalisation that motivates action. Analytics is suggested here as a game-changing partner for design. Working together as the foundation of the framework for competitive differentiation, they can deliver a powerful business value proposition.

ANALYTICS

Context

The new currency in business is data: information in raw or unorganized form (such as alphabets, numbers, or symbols) that refers to, or represents conditions, ideas, or objects (BusinessDictionary.com, n.d.). As the quantity, speed of delivery, and types and formats of data increase, the value received from acquiring, managing, and exploiting it accelerates. And one thing can be counted on: As a millennial manager or professional, you will be challenged with an unceasing explosion of data as the "Internet of

Everything" comes online and things gain contextual awareness, increased processing power, and greater sensing abilities. Cisco, a major innovator of Internet infrastructure

> defines the "Internet of Everything" as bringing together people, process, data, and things to make networked connections more relevant and valuable than ever before—turning information into actions that create new capabilities, richer experiences, and unprecedented economic opportunity for businesses, individuals, and countries. (Evans, D., 2012, p. 1)

By 2020, 31 billion devices and 4 billion people will be interconnected. "Just one business, Walmart, currently collects 2.5 petabytes of data on their customers every hour, a petabyte being the equivalent of 20 million filing cabinets full of text documents" (McAfee & Brynjolfsson, 2012, p. 62). This big data consists of messages, updates, and images posted to social media networks; readings from sensors on equipment and machinery; global positioning sensor (GPS) signals from cell phones, streaming video, and much more (p. 65). This is all data that has not been widely available before, and you will be challenged to figure out how to leverage it to increase the fidelity of your personal and organizational business decision making. One answer to this challenge is the latest evolution in this technical field known as "advanced analytics."

"Companies that have moved beyond data mining and management to quantitative and qualitative analysis and visualization techniques integral to the corporate strategy are companies that are exploiting knowledge for competitive advantage" (Grace, 2009, p. 10). The rise of advanced analytical tools and technologies from the backwaters of decision support to enterprise-level business strategies is a catalyst for a fundamental shift in business priorities. The dynamic emergence of analytics as a way to tame the wilds of "big data" has taken center stage in the business press. The hype around big data will lessen, but analytics will remain a strategic imperative and will come to characterize the data-intensive postmodern knowledge era.

The quandary of how to extract value from data is driving a management revolution that is fundamentally changing "long-standing ideas about the value of experience, the nature of expertise, and the practice of management" (McAfee & Brynjolfsson, 2012, p. 62). "We are on the cusp of an analytics revolution that may well transform how organizations are managed [as well as the] societies in which they operate" (Kiron, Ferguson, & Prentice, 2013, p. 2). Decision makers across the globe who have been highly compensated for their intuitive and experience-based business knowledge are engaged in a serious paradigm shift as they learn to accept as input to their decision-making process evidence that is compiled by data experts who study patterns in vast datasets and translate those patterns into business insight. This will drive serious changes to organizational cultures and structures as analytics experts are integrated into existing teams or formed up into centers of excellence that serve entire enterprises. The silos that today keep IT talent and business knowledge separated will be bridged by integrative analytics functions that require deep business knowledge to frame the right questions, and deep technical knowledge to build the models that will operate on the data, which, as well, requires IT expertise to acquire and manage the data and then build compelling visualizations to maximize insights.

As more and more activity is digitized, equipment and storage costs become ever cheaper, and technology tools that operate to explore and mine valuable insights become easier to use, a new era dawns where business insights can be gleaned on any topic of interest. Business and technology management are at a point of convergence. Technology has become integral to business, in either its products and services or its internal processes or both. A data-oriented culture, which is a "pattern of behaviors and practices by a group of people who share a belief that having, understanding, and using certain kinds of data and information plays a critical role in the success of the organization" (Kiron et al., 2013, p. 18), is becoming the norm.

Value and Purpose

You can get an edge by recognizing this change and seizing this opportunity by blending data-savvy people into your organization and having them mentor you in how to use data to enhance your daily decisions. That you can realize a competitive advantage by embracing data-driven decision making is supported by recent research: "Companies in the top third of their industry in their use of data-driven decision making were 5% more productive and [their businesses] were 6% more profitable than their competitors" (McAfee & Brynjolfsson, 2012, p. 64).

These are metrics that should determine your learning agenda going forward. You can't be considered someone with executive potential in today's business climate unless you can talk analytics. You must also be capable of digesting and leveraging the data analysis results that are fed through multiple management channels to enable data-driven decisions. This is a minimal success formula. To have an edge over your competition, you must be creating an analytics center of excellence within your organization that focuses on a critical data domain that not only informs your decision-making, but provides residual value to the enterprise as a whole. If you are not augmenting your current staff with analytically savvy people then you are falling behind.

In a 2012 survey by *MIT Sloan Management Review* (Kiron et al., 2013) of over 2,500 respondents in over 24 industries, the results showed that companies were exploiting analytics to gain a competitive edge. A further finding was that some companies were extending this advantage and leveraging analytics to enable a culture of innovation across their corporate ecosystems. These "Analytical Innovators" demonstrated three key characteristics:

- A widely shared belief that data is a core asset that can be used to enhance operations, customer service, marketing and strategy
- More effective use of more data for faster results
- Support for analytics by senior managers who embrace new ideas and are willing to shift power and resources to those who make data-driven decisions (Kiron et al., 2013, p. 2).

If companies are gaining a competitive edge from analytics, then the same applies to you. The success factors are well documented. "The digitization of the economy is one of the most critical issues of our time. The broad use of analytics is an important factor in the development of the digital economy" (Kiron et al., 2013, p. 3). But it is important that you realize where you need to focus. It's about helping you and your business colleagues make better decisions—this is where the business payoff comes, and that's why "organizations get [on average] $10.66 of value for every $1.00 invested in analytics" (Evans, D., 2012, p. 1). Now that's an ROI that will make your balance sheet shine!

"Using Big Data to decide on the basis of evidence rather than intuition" (McAfee & Brynjolfsson, 2012, p. 65) was born in companies like Google and Amazon, but the potential for competitive advantage for all businesses is real. Right now, analytics are generally separate from operational processes and require management initiative to enable extraction of the business value. But product development innovators will be looking to embed analytics in every conceivable way to give customers real-time information to guide their decision making. Customer preference data will be automatically captured and fed back to shape the products of the future. Now is the time to get ahead of this revolution and foster a culture within your locus of control that will enable ubiquitous data-supported judgments. Leadership in an era of big data is more critical than ever, as the choices available will multiply and selecting the right direction in a sea of directional information requires a knowledgeable hand on the tiller. Those who aspire to occupy executive suites must become better at asking the right questions, as that is what drives the analytical engine. But finding people who can operate that engine will not be easy.

Analytical talent and technological know-how, the coupling of statistics and computer science, are a critical combination of skills that is just now showing up on the radar of technical schools and universities. It is a new profession, but with some fairly old roots. It has been named "data science" to differentiate this new branch of analytics from statistics and business intelligence (BI), which is basically a query and reporting function that operates on structured database information. John W. Tukey is credited with first stating that "data analysis is intrinsically an empirical science" (as cited in Press, 2013) and arguing that "more emphasis needed to be placed on using data to suggest hypotheses to test and that exploratory and confirmatory data analyses can and should proceed side by side (as cited in Press, 2013). This focus on the front end of problem solving, the focus on hypothesis forming and discovery, creates a perfect partnership with design methods. Design practice hones the skills needed for excellence in analytics.

In 1977, the International Association for Statistical Computing (IASC) established its mission "to link traditional statistical methodology, modern computer technology, and the knowledge of domain experts in order to convert data into information and knowledge" (Press, 2013, p. 1). Data science, which

> incorporates varying elements and builds on techniques and theories from many fields, including mathematics, statistics, data engineering, pattern recognition and learning, advanced computing, visualization, uncertainty modeling, data warehousing, and high performance computing with the goal of extracting meaning from data and creating data products (Data Science, n.d.) was born from these origins. Pattern recognition and learning is another skill set that is essential to both design and analytics, and modeling, which is fundamental to analytics work, is also the preferred way to communicate design ideas.

The discipline that has emerged from the science of data as most valuable in the context of gaining competitive advantage is predictive analytics—"technology that learns from experience (data) to predict the future behavior of individuals in order to drive better decisions" (Siegel, 2013, p. 11). Clearly, design methods and the science of data analysis share a similar goal: enhanced organizational decision making. There are other types of analytics: descriptive, which describe the past, but don't provide any insight into why things happened the way they did; diagnostic, which provides hindsight about why something happened; and prescriptive, which "enables decision-making to not only look into the future . . . and see opportunities (and issues) that are potentially out there, but . . . also presents the best course of action to take advantage of that foresight in a timely manner" (Basu, 2013, p. 8).

"We live in a predictive society . . . [and] the best way to prosper in it is to understand the objectives, techniques, and limits of predictive models" (Davenport & Dyché, 2013, p. xiv). Predictive models yield information about "things that haven't happened yet" (Siegel, 2013, p. xv),

> using what we *do* know—in the form of data—to place increasingly accurate odds on what's coming next [blending] the best of math and technology,

systematically tweaking [the data] until [we derive insights] that peer right through the previously impenetrable barrier between today and tomorrow. (Siegel, 2013, p. xvii)

Design methods provide practice in this type of possibility thinking or the use of abductive logic, as described previously, which hones prescriptive and predictive analytics skills. And the person who excels in these skills in today's professional ranks is known as a data scientist, the new "rock star" in the business firmament.

Data scientists' most basic, universal skill is the ability to write code...[a] more enduring [skill] needed for data scientists to communicate in language that all their stakeholders understand—and to demonstrate the special skills involved in storytelling with data, whether verbally, visually, or—ideally— both. (Davenport & Patil, 2013, p. 73)

The profession of data science was created to design the algorithmic models that learn by

finding patterns that appear not only in the data at hand, but in general, so that what is learned will hold true in new situations never yet encountered [that is] the ability to generalize—the magic bullet of predictive analytics. (Siegel, 2013, p. 15)

Tom Davenport, perhaps the chief guru of the analytics revolution, points to an evolutionary shift in the analytics world, which he calls "Analytics 3.0" (Davenport & Patil, 2013, p. 72), blending "big data and traditional analytics that yield insights and offerings with speed and impact" (p. 27). This evolution, he argues, enables the spread of analytics to the masses and enables firms "in any industry [to] participate in the data-driven economy" (p. 27). Davenport also predicts increased emphasis on prescriptive analytics, which uses models to specify optimal behaviors and actions. Prescriptive analytics can be employed as "a means of embedding analytics into key processes and [influencing] employee behaviors [providing] a high level of execution [and changing fundamentally] the roles of front-line employees and their relationships with supervisors" (Davenport & Patil, 2013, p. 70).

Analytics in Your Organization

Clearly, if you aren't preparing for this change by growing and hiring this talent in your organization today, someone else will and they will have the edge that should have been yours.

As markets shift to what has been called "the intelligent economy, " or the convergence of intelligent devices, social networking, pervasive broadband networking, and analytics, there is a significant decrease in the ability of managers to rely effectively only on experience or intuition to make decisions. The old cause-and-effect mental models are becoming less relevant, while the demand to respond faster and with greater insight to ongoing internal and external events based on facts is increasing (Vesset & Morris, 2011, p. 1)

Look around you, analytics are everywhere: in that antitheft device that establishes your identity, that ad for an iPhone you received, that vibrating seat beneath you that returns your attention to where you are driving, the insurance premium adjustment you received based on your driving habits (Siegel, 2013, p. 219), and there will only be more. But you still have time to get ahead of the curve on this, if you move now. According to the recent MIT Sloan Research Report,

Analytics are mired in "automating the existing" rather than innovating a brighter performance future. It is a culture problem that will not be mitigated until a real leadership change occurs. . . . Analytics practitioners tend to be focused more on day-to-day operational use of analytics, as opposed to using it to drive innovation and change the business. (cited in Kiron et al., 2013, p. 15)

You could be that leadership change that "teaches the enterprise to behave differently with data and move from a transactional to an insight mind-set" (Kiron et al., 2013, p. 15). Blending design methods and advanced analytical capabilities will enable your organization to accelerate decision making to real time and increasing insights into customer intentions—for example, by predicting customer behavior so your products and services are positioned to anticipate customer needs. It's not magic, its analytics!

But don't underestimate the overhead that necessarily comes with building up an effective analytics capability inside your organization. You will need to seed your team with the right talent, and some of that talent base is hard to find because it is really just forming up. Data analytics requires expertise in data science, which "is a four-headed organism with focuses on business, data, analytics and narrative . . . [requiring] someone who knows how to extract meaning from and interpret data [as well as exceptional] persistence, statistics and software engineering skills" (Miller, 2013). Not easy to find and harder to attract without a whole system strategy such as we outline in this chapter.

Predictive and prescriptive analytics and advanced visualization technologies and techniques operate synergistically with design approaches. Together, they create a powerful pairing to drive a business's innovation engine. Managerial leaders and professionals of the future must be fluent in the data sciences both to explore and exploit the insights that the constantly growing sea of information yields but also to effectively work with their teams. Data scientists, analysts, and information architects will be on recruiting agendas in steadily increasing numbers over the coming years. You will need to know how to talk to them, select out the winners, attract them to your team, and leverage the insights they will deliver to enhance your business decisions. But in order to keep them and replicate their skills, you'll undoubtedly need a serious organizational culture transformation.

Interestingly, a culture conducive to analytics shares many of the same attributes as a culture that values innovation. Analytics work, because of its emergent properties, provides an excellent focal point for inventions, intellectual property, and in general innovations that deliver real business value. Fostering a culture of innovation in your organization is a natural fit in an organizational context where design principles are being practiced, design thinking methods and tools are being used as a part of daily work, and advancements in analytics are exploring and exploiting possibility thinking.

INNOVATION

Context

Innovation has always been the Holy Grail for business, but like the Grail, innovation appears elusive, mysterious, and ill-defined. It has been the stuff of legends and its secrets have been held closely—adding to its mystical persona, encouraging the notion that only a few are worthy of the quest. But in this new millennium this notion is being turned on its head. The quest for innovation is radically transforming the design of organizational systems now, and will be an ongoing reality in the coming decade. Innovation is breaking out of its R & D box, and an innovation movement is in the forming stage. You have a golden opportunity now to explore the subject of innovation, sensitize yourself to the spirit of innovation, and become an evangelist for innovation in your organization.

Innovation and Design

Innovation can be understood as the process of implementing a new idea into a product or service that has business value. Making the process repeatable with a consistent or consistently escalating value proposition is the challenge of today. Vishal Sikka from SAP has explored the question of how some companies manage to produce products that are "beautiful, appealing, intuitive, that connect with us emotionally...are well engineered, well designed, create economic value" (Sikka, 2012, p. 1) and how they manage to do it over and over again; how they manage to build innovation excellence into their culture. He argued that great innovators demonstrate deep understanding in three primary areas: (1) desirability—achieved through empathy with the end users of the product; their emotions and the psychology that influences their actions, and how they experience the product; (2) feasibility—achieved by being close to the details; what the product does, why and how it does it, and what it does not do; and (3) viability—the value created, the commercial implications, the economics. This recipe, argues Sikka, tells us "why organizational silos is such a bad idea… they reduce our proximity to customers, products, and value" (Sikka, 2012, p. 4). Silos are the enemy of collaboration and prevent the formation of a culture focused on design, analytics, and innovation, as these competencies thrive in a multi-perspective and multidisciplinary environment of open information sharing. The principles and practices of design thinking put empathy first, followed by iterative exploration and discovery of a shared point of view before any requirements are gathered—all essential innovation capacities.

> True innovation is about discovery, learning, and analysis rather than building on past knowledge and success. True innovation is unusual and requires a different approach. Innovation is the only activity in a business where the people involved are amateurs, because we spend so little time in most businesses working on discovering new needs, learning about customers and markets, and thinking deeply about the implications from discovery and learning. (Phillips, 2013, pp. 1–2)

MIT in 2012 identified design "[as] a driver of innovation and productivity" (Hobday, Boddington, & Grantham, 2011b, p. 18) and named design as a "core creative industrial and economic activity" (Hobday et al., 2011a, p. 5). Clearly, what they were predicting was an increasing convergence of design and innovation research and practice. MIT also drew together and recognized the interdependent relationship between technological innovation and organizational innovation (p. 6) and argued that "the concept of design, like innovation, has recently broadened to include non-technical areas of human activity, such as policy, organization, and social issues" (p. 6). When design is not viewed narrowly as a problem-solving approach, but more importantly and holistically as a knowledge generation and integration activity (p. 18), design practices can be seen to enable innovation while also influencing change in organizational structures to better support the close and consistent collaboration required of design and innovation action. When one adds the trump card of technology innovation, especially in the area of data analytics, a transformation of management practices and organizational systems will necessarily result.

Fostering Innovation

An open attitude toward learning cannot be created overnight. A culture that engenders trust is built over time. Using an innovation agenda to start that journey is a decision that, if managed appropriately, can both deliver business dividends from fostering creativity and reap other benefits from improvements; for example, in employee morale and productivity. To sustain an innovation agenda, however, requires clearly defined rewards and recognitions; formal and informal organizing principles that encourage and enable deep collaboration; processes that facilitate information sharing; design methods and tools that get embedded into daily work routines; as well as vocal, and perhaps more important, budgetary leadership support. Sustainable innovation cannot be a linear exercise, but must be designed from the start to be dynamic and focused on collective participation and contribution. Cross-organizational and cross-disciplinary collaboration has to be enabled and encouraged, and that will necessarily drive change into traditional organizational boundaries, like budgets; in short, what is required is whole system change.

The basic framework of innovation begins with research—"exploration of new frontiers, punctuated by occasional flashes of insight that lead to exciting new discoveries" (Chesbrough, 2003, p. 31). Design thinking methods provide a road map for guiding this research toward viable innovation targets, such as analytics as described previously. "Design can permeate every aspect of product and service development, naturally reaching into the innovation process.... The open-mindedness developed by radical teamwork and the exploratory procedures found in design are prerequisites for the innovators of the future" (Kim & Baek, 2011, p. 82). But there is nothing that feeds the innovation engine more than a continual flood of ideas and the excitement that comes from that idea generation. Leveraging the discipline of design methods provides the structure for how to focus ideas on a particular target or set of targets.

> Managers frequently believe that letting chaos reign can unleash their company's innovative energy. Removing boundaries, the logic goes helps managers spot or create innovative growth businesses. Yet, companies often come to realize that having a blank slate can make it surprisingly hard for managers. (Anthony, Johnson, & Sinfield, 2008, p. 4)

It is important to channel and manage employees' ideas so they can be processed through the innovation pipeline. Design methods set constraints and guide innovation, which "paradoxically... can be liberating" (Anthony et al., 2008, p. 1). Training in recognizing and qualifying disruptive ideas (p. 1) based on a set of clear criteria is one area that is underappreciated and therefore underleveraged. When you structure your innovation initiative from a whole system perspective these relatively low-cost enablers can make a big difference and also allow everyone to feel a part of the innovation challenge. But there is a caution: Releasing the spirit of innovation, even within a set of well-defined boundaries and especially when employing open-source technologies/social media, you can expect an overflow of ideas that will fill the pipeline, at first. So before you open the floodgates on any innovation initiative make sure it is supported by a well-oiled and well resourced structure and process that evaluates and qualifies ideas, rewards thoughtful inputs, follows all ideas through, and provides feedback on outcomes to the employee who provided the insight. Without follow-up, new idea submittal will wither quickly and it will be difficult to reclaim enthusiasm once lost. Innovation, when understood from a systems perspective, will drive organizational culture change. How you manage it will determine whether the change advances the competitive advantage of your business or sets it back.

There is nothing magical about fostering a culture of innovation—it is hard work and the biggest obstacle isn't budgets, technologies, processes, or tools—although all of these are vitally important. The biggest obstacle to innovation is individual motivation—having a compelling target (like analytics) to attract your more left-brained individuals and channeled through the right-brain engagement of design methods is a structure that can drive the behavior you seek. Addressing motivation up front is the way to sustain the momentum needed to keep the whole system continually renewing the natural human ability to innovate.

THE WHOLE SYSTEM

"The year to year viability of a company depends on its ability to innovate" (Nagji & Tuff, 2012, p. 66) but without a holistic strategy that supports "total innovation" (Nagji & Tuff, 2012, p. 66), fragmented initiatives will dilute even the best ideas and end up wasting scarce corporate resources. "Overcoming the systemic challenges of collaborative innovation and applied human creativity" (Cooperrider, 2010, p. 26) is a whole system issue but one that offers truly breakthrough payback, because "to value innovation in systemic design terms is indeed to value one of the most abundant, renewable resources we can draw on" (Cooperrider, 2010, p. 26).

"Design thinking is a creative process based around the construction of ideas [and intends to] eliminate the fear of failure and encourages maximum input and participation" (Celaschi et al., 2012, p. 8). A system, whether political, economic, social, technological, or educational, is structured to enable behaviors that serve purpose and goals. Perhaps more than any other time in history, business organizations have become the domain where our institutional attention is focused and where technological capabilities are joining other core competencies to fundamentally change the business landscape. What has our attention today is that "modern corporations [arguably have become] the most powerful institutions... [and] in many ways they govern modern life" (Kelly & White, 2007, p. 3). Discontinuous changes that rattle global business, therefore, affect the whole system.

Designing the Business Ecosystem

Increasingly, we are expecting our corporations to behave in ways that their traditional siloed organizational structures disallow. We need to recognize this mismatch and intentionally design our business ecosystem to serve all stakeholders, not just stockholders, but also to enable corporate cultures to facilitate business model reinvention realities and accommodate the changed requirements for work structure and behavior. "Corporate design is the missing business and public policy issue of the day" (Kelly & White, 2007, p. 3) and we need to look at this subject from multiple perspectives holistically, through the sensibilities and capacities of a designer.

If the Internet has taught us anything, it has shown us a model for the complete interconnected nature of existence; although we have not yet tackled the overall lack of quality in the intercourse that we have enabled with this technology, nevertheless global interconnectedness is, at least technologically, a reality. There is a growing awareness that we cannot continue discussing issues as if they were separate and unrelated to other issues within a contextual space. "We face today a historical moment when a fragmented, reactive approach to corporate responsibility must give way to a systemic and structural approach commensurate with the expanding economic, ecological, and social footprint of the modern corporation" (Kelly & White, 2007, p. 3). Integrating design thinking and design methods into organizational forms offers a structure for work that is optimized for people, for enabling their creativity. Human-centered design and its core competency, interaction design, are the specific capabilities that can catalyze "desirable relationships both inside and outside the organization, using product development as the vehicle to shape these relationships and ultimately the organization itself" (Junginger, 2004, p. 5).

> Thriving in the midst of today's frenetic pace of change requires a new set of approaches and tools. Incremental change may have been enough at the end of an industrial era marked by "me-too" products and services, process re-engineering, best practices, benchmarks, and continuous improvement. (Kaplan, 2012, p. xiii)

What is required today is consciousness-shifting mental model transformation that will enable the innovation explosion that both business executives and knowledge workers are asking for. Only a revolution in organizational cultures will be able to deliver this level of change. Businesses are turning the corner on the realization that they must learn to fully engage the customer in the design of products and services. "The ability to focus on customers is viewed as the top ranked factor for developing an innovation culture," according to a Human Resource Institute survey of over 1,300 global respondents in 2005 (Jamrog, Vickers, & Bear, 2006, p. 12). The realization is dawning on business enterprises that fully engaging their employees by creating authentic human-centered organizations is the path to developing the level of commitment required for employees to release their innate abilities to innovate in all areas of corporate work, including that most vital focus on the customer. Adopting design as a discipline works for this task.

Since Drucker identified "innovation and entrepreneurship as part of the executive's job" (1985, p. vii), business executives have been looking for a "how to" guide to innovation but not finding the answers in traditional MBA curricula. The new "d. schools" are the places where executives can now learn to "leverage design to enable their organizations to innovate [where they can learn] how to do business with design [and] create competitive advantage through design" (Verganti, 2009, p. 25). Over the next decade, a master's degree in Business Design, or a master's degree in Fine Arts (Pink, 2005, p. 54), will emerge to challenge the traditional master's in Business Administration (Grace & Graen, 2014). "The direction of academic change toward entrepreneurial sciences is constant" (Etzkowitz, 1999, p. 203). There is "significant pressure on universities to pursue greater commercialization efforts" (Schoen, Mason, Kline, & Bunch, 2005, p. 3) as commercialization will increasingly require understanding what people are trying to achieve when they buy products. This vital piece of the puzzle of customer meaning-making and industrial market-shaping is already emerging in the curricula of business design education. This is a trend you should pay attention to.

Higher Order Thinking and Organizing

Fred Collopy and his colleagues at the Weatherhead School of Management, Case Western Reserve University, introduced design as a management discipline in 2004 (Boland & Collopy, 2004). Over the last decade, Dr. Collopy has continued to contribute to the thinking in this area and his university is one of the increasing numbers of graduate schools that have placed design studies on a par with traditional MBA programs. In a recent article, Dr. Collopy discusses the synergies between more right brain thinking, like design, and the need to marry that up with more left brain analytical type thinking.

Collopy writes,

> one of the most compelling arguments for design thinking is that when people engage design competencies alongside their analytic ones, they function as whole human beings. . . . we must take seriously the challenge of understanding how these two very different ways of being can be brought into an accord that will function both at an individual level and to change our culture. . . . we must balance our analytic understandings with our intuitions. (Collopy, 2010, p. 1)

Uniting design and analytics leverages their complementary thought processes into a more holistic thought paradigm. Conceptually, "the left hemisphere participates in the analysis of information . . . the right hemisphere is specialized for synthesis and is particularly good at putting isolated elements together to perceive things as a whole [and] both are essential to human reasoning (Pink, 2005, p. 22). As suggested in Figure 2.2, the combining of linear and non-linear thought paradigms applies the power of

Figure 2.2 Whole system design. © 2014 Miriam Grace.

holistic thought to an innovation context. The triad of design, analytics, and innovation, when they are well managed and operated as a whole system change engine, have the potential to shift paradigms and directly and positively grow an organization system. The synergies among the operating elements create a self-reinforcing loop—a virtuous cycle that can be an engine for transformational change within your business ecosystem.

Having designers in your organization or project ensures a focus on empathy, shared point of view, multidisciplinary collaboration, prototyping, iteration, and feedback. However, as the Father of Industrial Design, Raymond Loewy (1893–1986), has often been quoted, "design is now too important to be left to designers" (Brown, 2009, p. 37). Increasing your awareness of the role design can play in the innovation value chain (Hansen & Birkinshaw, 2007) will allow you to work toward an innovation-friendly environment in your team that can be a model for your enterprise for how to create sustainable competitive differentiation.

The reality is that product innovation is easily replicated, but a design culture [that produces] process and service innovations [is] much harder to copy.... Product innovation gives less than three months' competitive advantage. Process innovation gives at least 12 months competitive advantage (Mehta, 2006, p. 3). Enabling a self-reinforcing process, however, becomes an enterprise asset that allows replication of innovation-generative behaviors and stretches out the length of time innovations deliver business value.

In addition to impacting internal structures, external affiliations or network alliances (Graen, 2013) with "interpreters" (Verganti, 2009, p. 144) from universities, industry groups, professional organizations, and think tanks will routinely contribute to the visioning of the team to create new meanings and markets for goods and services. The team will be operating with the very latest global information and will integrate it into their own data analytics and visualization engines to deliver real-time knowledge and insights about their company, its customers, its competition, and its adjacencies in the context of world markets. Potential threats to their market position can thus be uncovered and proactively addressed.

"The goal [of design] is to understand what people perceive, what their individual perspectives or frames of reference are and what meanings follow from their different points of view" (Buchanan, 2001, p. 80). This allows for emergent solutions that address the moving present (Buchanan, 2001). From this perspective, it is easy to recognize the millennial management paradigm where "whole system" decisions must be made that require an actual unfreezing and transitional change in organizational culture (Lewin, 1948) but the transition now must be to a much more fluid and adaptive organizational paradigm and structure than was previously possible. In such a dynamic environment as the future promises, the most seductive business questions can be asked: What if? What might happen? What is possible?

How risky is it? What should happen? What can I optimize? Designers will readily recognize many of these questions as classic examples of the application of abductive logic, the possibilities-based reasoning approach characteristic of design. Data scientists will recognize them as the territory of advanced analytics. The marriage of design and analytics is inevitable.

Roger Martin, founder of the Rotman MBA program, put it this way in a recent thought interview: "The fact is that if you're going to be special in today's world, you have to go to the next higher-order level of thought" (Christensen, 2013, p. 10). Looking through a whole system lens and finding ways to get everyone engaged in the adventure is a next higher-order level of thought. The fact that design thinking is emergent in business contexts in parallel to the ascendance of synergistic technological innovations like advanced analytics is not pure coincidence. The technology is simply another recognition that we have arrived at a time in our history when the magic of prescience is just one more example of our abilities catching up with our imaginations.

CONCLUSION

This chapter has provided you the insight to understand how design and analytics support one another and provide methods and tools that can drive innovation. Integrating the principles and practices of design and the mind-set and methods of design thinking into daily work supports the cultural requirements for building an effective analytics center of excellence where breakthrough innovation becomes the shared ambition. Guiding the evolution of these three interrelated elements so they operate synergistically will contribute, over time, to positive organizational change via creative growth of the whole system.

This chapter has both predicted and supported the argument that the future of organizational systems will be driven by the need to collaborate in order to continuously innovate. Information flows will be consumed by ubiquitous analytics that add value by identifying opportunities, predicting outcomes, and monitoring patterns and trends to inform real-time decisions. Millennial managers and professionals will need to be fluent in design principles and practices in order to hire employees with the rich diversity of disciplines, perspectives, skills, and talents required to keep the idea engine stoked and to have the know-how to operationalize those ideas into compelling products and services. Deep customer engagement and a diversity of network alliances will be hallmarks of a successful organization. The millennial manager and professional will have developed deep empathy for the customer and other stakeholders, will understand what drives their customers' businesses, and will consider them as part of the extended design team. Universities will need to augment their executive learning curricula with these advancements in understanding how to engage the whole system in an organizational change agenda.

Now, and increasingly in the future, the ability of an organization to make innovation a core competency and to continuously expand its innovation capacity creates a counterbalance to the complexity of a global marketplace where emergent competition is constantly challenging market share. The discipline of design and the mind-set of design thinking have been proven to successfully address the all-important Why? question, to ensure a business has the right focus. Design methods answer the How? question by providing guidance and discipline to ensure innovation activities deliver business value. Analytics is suggested as a focus for answering the What? question relative to an innovation agenda because it has an established ROI, encourages a mental model that can encompass the vastness of today's information landscapes, and likely will be readily funded. The business value of analytics is realized in its ability to identify, manage, and mine data to expand business insights and inform both strategic and tactical decision making. Advanced analytics practice, and the information management revolution that accompanies it, offer a green field for experimentation and discovery, and perfect partnership for operationalizing of the methods and tools of design thinking. Bringing design, analytics, and innovation into your consciousness enhances your sensitivity to opportunities and insights where any one or all three capabilities might bring business value. As you continue to invest in such a holistic mindset, you will naturally be led toward positive change. That's just the way the whole system works.

REFERENCES

Anthony, S. D., Johnson, M. W., & Sinfield, J. V. (2008, Winter). Institutionalizing innovation. *MIT Sloan Management Review*. Retrieved December 6, 2013 from http://sloanreview.mit.edu/article/institutionalizing-innovation/

Basu, A. (2013, March/April). Five pillars of prescriptive analytics success. *Analytics Magazine,* 8–12.

Bolman, L. G., & Deal, T. E. (1997). *Reframing organizations: Artistry, choice and leadership.* San Francisco, CA: Jossey-Bass.

Boland, R. J., & Collopy, F. (2004). *Managing as designing.* F. Collopy & R. Boland, Jr. (Eds.). Redwood City, CA: Stanford University Press.

Brown, T. (2009). *Change by design: How design thinking transforms organizations and inspires innovation.* New York: Harper Collins.

Buchanan, R. (2001). Children of the moving present: The ecology of culture and the search for causes in design. *Design Issues, 17*(1), 67–84. doi:10.1162/07479360152103840

BusinessDictionary.com (n.d.). *Data.* Retrieved December 10, 2013, from www.businessdictionary.com/definition/data.html

Celaschi, F., Celi, M., & Garcia, L. M. (2012). The extended value of design: An advanced design perspective. *Design Management Journal, 6*(1), pp. 6–15.

Chesbrough, H. W. (2003). *Open innovation: The new imperative for creating and profiting from technology.* Boston: Harvard Business School Press.

Christensen, K. (2013). Thought leader interview: Roger Martin. *Rotman Management,* Fall 2013, 10–15.

Collopy, F. (2010, June 18–19). *Firing on all eight cylinders.* Paper presented at Convergence: Managing + Designing, Cleveland, OH. Cleveland, OH: Weatherhead School of Management, Case Western Reserve University.

Cooperrider, D. (2010). Managing-as-designing in an era of massive innovation: A call for design-inspired corporate citizenship. *Journal of Corporate Citizenship, 37,* 24–33.

Data Science. (n.d.). In *Wikipedia.* Retrieved December 12, 2013, from http://en.wikipedia.org/wiki/Data_science

Davenport, T. H., & Dyché, J. (2013). *Big data in big companies* (Tech. Rep.). Cary, NC: SAS Institute Press.

Davenport, T. H., & Harris, J. G. (2007). *Competing on analytics.* Boston: Harvard Business Press.

Davenport, T. H., & Patil, D. J. (2013). Data scientist: The sexiest job of the 21st century. *Harvard Business Review, 90*(10), 70–76.

Etzkowitz, H., (1999). Bridging the gap: The evolution of industry-university links in the United States. In L. M. Branscomb, F. Kodama, & R. Florid (Eds.), *Industrializing knowledge* (pp. 203–33). Cambridge, MA: The MIT Press.

Evans, D. (2012, November 7). How the internet of everything will change the world...for the better. *Cisco Blogs,* Retrieved December 2, 2013, from http://blogs.cisco.com/ioe/how-the-internet-of-everything-will-change-the-world-for-the-better-infographic/

Grace, M. (2009). What is a game-changing design? In G. Graen & J. Graen (Eds.), *Predator's game-changing designs: Research-based tools. LMX leadership: Vol. 7. LMX leadership: The series* (pp. 1–18). Charlotte, NC: Information Age Publishing.

Grace, M., & Graen, G. (2014). *What if we designed an MBA for the future?* Unpublished manuscript.

Graen, G. B. (2013). The missing link in network dynamics. *The Oxford handbook of leadership* (pp. 359–375). Michael G. Rumsey (Ed.). London, UK: Oxford University Press.

Hansen, M. T., & Birkinshaw, J. (2007). The innovation value chain. *Harvard Business Review, 85*(6), 121–130.

Hobday, M., Boddington, A., & Grantham, A. (2011a). An innovation perspective on design, part 1. *Design Issues, 27*(4), 5–15. doi:10.1162/DESI_a_00101

Hobday, M., Boddington, A., & Grantham, A. (2011b). An innovation perspective on design, part 2. *Design Issues, 28*(1) 18–29. Cambridge, MA: MIT Press. doi:10.1162/DESI_a_00137

Jamrog, J., Vickers, M., & Bear, D. (2006). Building and sustaining a culture that supports innovation. *Human Resource Planning, 29*(3), 9–19.

Junginger, S. (2004). *A different role for human-centered design within the organization.* [PDF file http://ead.verhaag.net/fullpapers/ead06_id197_2.pdf]. Retrieved November 26, 2013, from http://ead.verhaag.net/00.php

Kaplan, S. (2012). *Business model innovation factory: How to stay relevant when the world is changing.* Hoboken, NJ: John Wiley & Sons.

Kelly, M., & White, A. (2007). *Corporate design: The missing business and public policy issue of our time.* Boston, MA: Tellus Institute.

Kim, B. Y., & Baek, J. H. (2011). Leading the market with design thinking and sensibility. *Design Management Review, 22*(3), 80–89.

Kiron, D., Ferguson, R. B., & Prentice, P. K. (2013). From value to vision: Reimagining the Possible with data analytics. (Research Report of *MIT Sloan Management Review*). Cambridge, MA: MIT Press.

Lekach, M. (2013). Know your design history: The Bauhaus movement. Retrieved December 10, 2013, from http://99designs.com/designer-blog/2013/08/15/know-your-design-history-the-bauhaus-movement/.

Lewin, K. (1948). The complete social scientist. In T. M. Newcomb, & E. L. Hartley (Eds.), *Readings in social psychology* (pp. 330–341). New York: Henry Holt.

Martin, R. (2004). *The design of business: Why design thinking is the next competitive advantage.* Boston: Harvard Business Press.

McAfee, A., & Brynjolfsson, E. (2012). Big data: The management revolution. *Harvard Business Review, 90*(10), 60–68.

Mehta, M. (2006). Growth by design: How good design drives company growth. *Ivey Business Journal, 70*(3), 1–5.

Miller, S. (2013, November 22). Data science—A tale of two books. *Information management: How your business works.* Retrieved December 4, 2013, from http://www.information-management.com/blogs/data-science-a-tale-of-two-books-10025087-1.html?ET=informationmgmt:e9513:1873755a:&st=email&utm_source=editorial&utm_medium=email&utm_campaign=IM_Blogs_120413

Nagji, B., & Tuff, G. (2012). Managing your innovation portfolio. *Harvard Business Review, 90*(5), 66–74.

Phillips, J. (2013, February 27). Why the front end of innovation is different. Innovation excellence. Retrieved December 5, 2013, from http://www.innovationexcellence.com/blog/2013/02/27/why-the-front-end-of-innovation-is-different/

Pink, D. H. (2005). *A whole new mind: Why right-brainers will rule the future.* New York: Riverhead Books.

Press, G. (2013, May 28). *A very short history of data science.* Retrieved May 28, 2013, from http://www.forbes.com/sites/gilpress/2013/05/28/a-very-short-history-of-data-science/

Schoen, J., Mason, T. W., Kline, W. A., & Bunch, R. M. (2005). The innovation cycle: A new model and case study for the invention to innovation process. *Engineering Management Journal, 17*(3), 3–10.

Siegel, E. (2013). *Predictive analytics: The power to predict who will click, buy, lie, or die.* Hoboken, NJ: John Wiley & Sons.

Sikka, V. (2012). Measure of an innovator: The innovator's index. *SAP Business Innovation.* Retrieved December 5, 2013, from http://blogs.sap.com/innovation/innovation/measure-of-an-innovator-the-innovators-index-01097

Silver, N. (2012). *The signal and the noise: Why so many predictions fail but some don't.* New York: The Penguin Group.

Verganti, R. (2009). *Design-driven innovation.* Boston, MA: Harvard Business School Press.

Vesset, D., & Morris, H. D. (2011). *The business value of predictive analytics* (IDC Report #229061). Framingham, MA: IDC Press.

Wasserman, A. (2013). *Design 3.0: How design grew from 'stuff' to sociotechnical systems and became too important to leave to designers.* Retrieved December 6, 2013, from http://designtoimprovelifeeducation.dk/da/content/design-30

CHAPTER 3

BUTTERFLY DESIGNS FOR Y-CAREERS

George Graen
The University of Illinois

ABSTRACT

Did you catch my blog about knowledge age firms being designed anew using twenty-first century principles for companywide, collaborative-design? If not, innovative and successful new ways of enabling human minds and bodies are emerging. New design methods are being applied to innovate all parts of organizations including physical layout, work flows and the right applications for operations, collaboration, strategy, and sustainability. Today, we are designing new wealth creating organizations that have continued unchanged since the industrial revolution. As you may have experienced, strategies, innovations in (1) child development methods; and (2) educational practices have resulted in wonderful young people. The new generations are uniquely prepared to explore and exploit the many opportunities of the knowledge age. I prophesize that the new drivers of "collaborative designing" and a generation of cyber prepared talent will combine to produce a sea change in our organizational behavior and human performance. Please prepare yourself by buying a ticket for a wonderful ride into an even better future. The procedures for purchasing a ticket to ride are opening your thinking and learning about the possibilities of healthy, attractive, engaging and rewarding ways to serve in firms.

Millennial Spring: Designing the Future of Organizations, pages 41–64
Copyright © 2014 by Information Age Publishing
All rights of reproduction in any form reserved.

BACKGROUND

Today, top management teams are worried about a new threat to business as usual. They fear the harsh realities of the perfect storm of a long economic recession, changes in people and changes in technology washing away employer rules and procedures designed for a bygone talent market. (PwC, 2013). Price-Waterhouse-Cooper, a giant service corporation, in response to sharply increasing resignation rates of their professionals before their second year hired the University of Southern California and the London Business School to find out what was broken. Why did under 30s professionals leave before two years of tenure?

This massive, big data self-study by PwC in 2012–2013 highlights the problem. After surveying over 40,000 employees and managers over a two-year investigation, the study investigators at USC and London Business School confirmed the reason that millennials (born 1980-2000) were accepting professional positions with PwC and resigning within two years. This research concluded that the under 30 professionals found present management practices dysfunctional for their intended careers. The PwC investigation concluded that times have changed and companies are advised to make their people operations more professionally functional for the under 30, because people have changed, technology has changed and the interaction between the two have changed.

What are the real problems? Over 30s complain that performance management systems only work for the in-group and not the out-group. Under 30s reject career management system and do not join the in-group. The differences between what LMX theory defines as in-group and out-group are several, (Graen, 2013). LMX theory is based on a mountain of scientific research over the last 30 plus years. It was constructed to describe and prescribe the team leadership motivating process of designing and creating the life of project teams. Professionals, who prove respected, trusted and team players are defined as the in-group and those who fail to pass these performance tests are the out-group in the unit. According to this body of research, the out-group has always been those dissatisfied with most components of the unit including the performance management system. The LMX leadership model prescribes that a new design for each unit should be developed that encourages all unit members to actively participate in an inclusive community construction process. The good news is that according to a current survey of CEOs this transformation has begun. In 2013, the World Economic Forum in Danos found over 75% of CEOs were planning to revise their management talent strategies that year. They were most concerned regarding the availability of key skills and the influence of this lack on both employees and clients in the future.

Today's advanced understanding and innovative processes for the design of more millennial (under 30s) friendly and sustainable organizations reveal how unfriendly and myopic are traditionally designed organizations. We now can do better. Under 30s often mention "stakeholders" when discussing the ideal beneficiaries of the American concept of corporation. In contrast, they protest that our organizations were built in the service of "shareholders." I argue that a new process called "collaborative-design mindset" is ready to build new under 30 compatible organizations. These new designs capitalize on the innovations flowing from the imagination of teams of respected and trusted professionals and managers employing the latest design technology (described in the preceding chapter). One strength of the new approach is that it strives to understand the efficiencies from many directions before the design is locked in place. This allows the realization of formerly unrealized economies. Beyond being more stakeholder friendly, these designs are more engaging for all employees, more attractive for customers and more wealth producing for all. It's time that we came aboard because this design spaceship is blasting off. Please don't get lost in thought at this choice point.

Looking Backward

Before blasting off, we need to consider where we came from to appreciate the new alternatives. Long before the United States existed, large companies were granted charters that permitted them to construct elaborate worker-machine systems employing large numbers of people. These charters from the owners allowed the bankers who held them to construct wealth-producing organizations. These companies were built to capitalize on the latest financial and engineering technology for profit and control. These drivers produced companies based on the latest specialization and division of labor for efficiency and a hierarchy of authority for control of the worker-machine systems (Merton, 1957).

The United States struggled to develop into the dream of welcoming immigration of those longing to be free from the problems of their native countries and seeking a better life for their families. Finally, in 1935, the U.S. Wagner Act gave workers the court enforced right to fair treatment from their employers. The law continues to seek more collaborative relations between the company directors (I call bankers) and their employees. It is not a pretty story, but great progress has been made to date. I predict that we are on the verge of dramatic improvements in the design of organizations.

Historically, large organizations really were not "designed" in the contemporary sense of the term. Rather, they were built to place people, as the multi-functional parts, in complex machine systems built using the engineering and finance ideas of the industrial revolutions. Historically, entrepreneurs who established successful enterprises were acquired by larger companies (I call bankers) and small companies were than cobbled into functional silos for ease of behavioral and financial control. While the shareholders focused designers on the creation and control of wealth, employee and customer needs often were after thoughts. Some CEOs of holding companies (banks) discovered that the way the power hierarchy was structured could be used to frighten employees to speed them on their jobs. Recently one CEO demanded that the bottom 10 percent of performing managers in his many companies be terminated each year. Clearly, the rationale of yesterday's modern organization was profit maximization and ease of financial control. One of their favorite strategies was to invest in the "star" companies, sell the "dogs," and milk the "cash cow." Employees were considered more costs than assets.

Power of Hierarchy

Bankers (company stock controllers) make terrible organizational designers. Their natural inclination leads them to act like their fathers or some other storied baron of wealth creation. They believe that the power hierarchy must be feared and obeyed or the owners may lose control of their dollars. The Tony Soprano character on the 2012 HBO TV series entitled "Sopranos" is a good exemplar. Tony is emotionally a frightened child with periodic fainting anxiety attacks. Nevertheless, Tony must act as an owner by copying his autocratic father, all the while hating his job as owner of the organization. His psychiatrist, with her counsel and medication, serves mainly as Tony's confessor. His employees fear him for his lethal temper and life or death authority. Like Tony, few entrepreneurs receive formal training before they start managing an organization. Much like most parents, they must rely on their experiences with their early care providers. This chapter is directed at the proper design of new organizations. This subject is vital to your future because organizations are given the power to engage you in profitable innovation, or burn you out quickly.

Organizational control of wealth creation can take many different forms of authority distribution. I argue that the new designs are better. The reasons for this are that new designs are based on deeper understanding of the nature of hierarchy and the common human tendencies enabling profitable sustainability in organizations (Pfeffer, 2013). Hierarchies of authority properly designed using a collaborative-design mindset aim to produce the company cultures described by Grace in her chapter. It is a culture that employs a power hierarchy to empower the drivers of the venture. In Toyota, the driver is the customers' appreciation for a satisfying experience in personal transportation. In contrast, Walmart's driver is the customers' appreciation for a continuous trustworthy and low priced source of obtaining groceries and other retail goods. Finally, Google's driver is the customer's appreciation for the creation of knowledge age applications called innovations. Hierarchy is a powerful tool and requires careful design and construction. It can enable or crush the driver of an organization. The story of Steve Jobs is a good example of an improper use of hierarchy to crush innovations in the name of short-term damage control (Isaacson, 2011). Jobs was terminated from the company that he founded only to be begged to later to take over a failing Apple. Many case studies of hostile takeovers end tragically. The problem for designers is to find the particular hierarchy and distribute authority in ways that validate the drivers of the company. To accomplish this, designers and professionals must discover the heart of the entrepreneur. The "what if" question to the designer becomes: What if we designed an organization with an authority structure that produces the most consistent image of its company culture? I would suggest that we begin with examining the design of Google Inc.

The rule is to structure authority and leadership flows to protect and support a company's purposes. For example, Google assigns authority to positions in the company that are concerned with the customer and the professional. It does not follow the traditional top-down pyramid structure of authority. Google assigns authority and enables leadership to the creative professionals in their innovation teams to support the most competent behavior by managers of their work environments. Other positions with specially designed authority and enabled leadership include the executive positions of enterprise architect in support of research and development, production, business development, sales, marketing and people systems. The authority and leadership networks need to be carefully researched to empower the drivers of the company. Finally, the use of authority and leadership distribution systems need to be monitored and coordinated as closely as profits and cash flows.

A Better Organization

Historically, people put up with some unnecessary discomfort at work from the old hierarchies for the sake of their families. Employee reactions to this disrespect sometimes was expressed in counterproductive actions. The question becomes how can we design organizations to avoid abuse and achieve more innovation, engagement and fun? Recent innovation in relevant design technology has shown that the design of the business organization can make even the bad boss better and talented professionals more innovative and fun (Bock, 2011). This new technology may serve our need to appropriately employ our rapidly descending internet generations before they have been burned out. The fortunate converging of new technology and new generations with different talent and expectations may turn out to be our opportunity to build better organizations and save entire generations including yours from future Tony Soprano-like power hierarchies.

The bottom line is that we need to design new structures and functions to restore our innovative edge (Hage, 2011). We lost 80 million jobs in the last decade and require new industries to replace them. Better workplace designs also are needed to facilitate the productive and engaging career transition of our new generations (Graen & Schiemann, 2013). These problems represent the challenges to the sustainability of our innovative edge. Fortunately, recent advances including a "collaborative-design mindset" promise to improve our chances to do what was impossible a decade ago (Boland & Collopy, 2004).

Our investigations of past holding companies revealed that the so-called "research" departments were staffed with myopic engineers who seldom looked up from their office computers. Research and design professionals seldom truly collaborated with their own teams and especially with those from other silos. They saw little reason for such interpersonal teamwork, because they were hired, trained, evaluated, promoted and earned raises and bonuses competing with their cohorts in their professional silo. Moreover, major disconnects are due to differences in cultures between and among the functional silos. Each functional area naturally develops its own culture within holding companies. In cases where the parent company had purchased several other firms, holding company bankers think that one overall culture should be commanded by whoever writes the paycheck. The problem today is that these holding companies are typically only a loose federation with little commitment to or from the subsidiaries. What seems to be missing is collaboration among cultures (Hui & Graen, 1997). Fortunately, contemporary designers have developed methods to build collaborative-design cultures. For example, Jahnke and Sköldberg describe in their chapter what happened when designers from a university were contracted to actively collaborate with manufacturing company managers on missions to produce more innovative and successful products. From a collaborative-design perspective, this experience opened new ways of thinking about the potential for innovation. Moreover, as these authors correctly point out, designing in the West and the East are culturally different. Their example is that the admired characteristics of an innovation are credited to the product in the West and to the person in the East. For example, Japan's national treasures are the creative artists not their products. We should note that products and services may fill the needs of many different cultures. These opportunities deserve to be thought about from the various cultural perspectives before the design is locked.

TWO CLASSIC CASES USING BUTTERFLY DESIGN

Two early case studies using collaborative mindset to build a butterfly design (Graen & Wakabayashi, 1994) company will be discussed next to show that an organization can be designed for a particular local market, but also prove effective in international markets. What is collaborative-design mindset 2.0? Fundamentally, it is a protocol incorporating Graen's work (2013a, b) on developing new ways to assemble and design project leadership teams to perform innovative improvements. This protocol directs talent to form teams using proper social networking tools such as Facebook, Twitter, the mobile web, smart phones and teleconferences. Teams form using these networking mechanisms and by negotiating *unique strategic alliances*. Let us consider my two exemplars, namely, the Toyota and Walmart creations.

Toyota's Butterfly Designs for America

My first experience with the collaborative-design mindset to integrate design with network alliance-making was the invention of a radically new firm I call the "butterfly design" in Toyota City, Japan. After the complete destruction of their family business, the Toyoda brothers asked themselves the key design science question. *What if we designed the perfect system for producing the highest quality and least wasteful world automobile?* Answering this question required them to discover the guiding principles of such a system. After a good deal of study of existing systems in the leading countries, they created the following parameters for what I call "butterfly" design. These new systems, much like monarch butterflies, require four different generations annually to survive and prosper and provide a critical service to sustainability. Toyota's new organizational structure and function included the following characteristics:

- Complete integration of "just in time" and "quality first" into all parts, all assembly, and all sales and service networks.
- Key suppliers financiers, distribution, sales, service, and labor union as alliance partners.
- Employees as fully engaged, multi-skilled members of the "Toyota clan."
- Team guardians as older siblings and coordinators.
- Transparent operations showing trust with the entire clan.

- Proactive problem finding and solving by overlapping quality groups.
- Overlapping team structure with team guardians as linking-pins.
- Teams at all levels housed in their own unique "cocoon" for security, creative problem solving and retention.
- Bottom-up flows of information and problem solving with focus on doing everything right the first time and continuous improvement (innovation).

This "butterfly" design for all manufacturing has become the *gold standard* worldwide. It works beautifully across national cultures. Many attempts to copy the "Toyota lean production system" piecemeal by consultants became fads in the West and eventually faded away. Those who understood design science were able, with instruction from Toyota Motors in many countries, to develop their own butterfly designs. Of course, the Japanese manufacturers were the first allowing them a head start in establishing plants in the West. This quantum leap forward in manufacturing globally must be credited to the courage and foresight of the innovative Toyoda family.

Mr. Sam's Butterfly Designs

At about the same time, Mr. Sam Walton, a small rural retailer returning from the war, asked his "what if" question. W*hat if we design the perfect system for delivering the world of retail at the lowest price to main street America?* (Graen & Graen, 2009). After a thorough investigation of existing systems, the Waltons discovered the relevant new parameters for their system. These were as follows.

- Concentrate on main street customers' repeat business.
- Integrate shareholders, suppliers, distribution systems, and retail stores into alliance networks to provide the lowest price to customers and profit sharing to all in the Walmart system.
- All associates are in the "Walmart clan" and share the strict ethical code of conduct.
- Each store is a separately managed cocoon with its own team's guardians, banners and song-filled social events.
- Suppliers share risks and rewards, and whole system is focused on collaborative-design and cutting-edge supply chain technology.

The Walmart Company is the largest company in the world in 2013. Its growth and service to the customer worldwide demonstrates the butterfly power of design thinking. This quantum leap forward must be credited to the courage and foresight of the Walton family. They employed the collaborative-design mindset to perfection. Next, we turn to the collaborative-design mindset studies that contributed to the most promising fundamental competence underlying today's team success in creating new strategic designs with cascading strategic executions (Graen, 2014).

RECENT HISTORY OF BUILDING A BETTER ORGANIZATION

The job of building a better organization captured the imagination of both scholars and practitioners with the passage of the National Labor Relations Act (NLRA, 1935) which gave employees the right to organize and negotiate legal contracts with their boss. For the first time, employees possessed court-enforced rights to "fair treatment." One of the first to present his theory was Chester Barnard (1938) based on his successes with Bell Telephone of New Jersey and the Rockefeller Foundation. Barnard, who was raised in a working class family, focused on the *hierarchy of authority* realizing that true cooperation from employees was problematic and rare. Thus, he emphasized the boss communication with employees as the key to becoming an effective company. Later, March and Simon's (1958) theory of the better organization prescribed the hierarchy of authority as depending on a balance between employee contributions and company inducements. Hackman and Oldham's job characteristic theory (1976) described the boss as the driver of jobs using the *hierarchy of authority* (managership). Leader member exchange (LMX) theory (Graen, 1976; Graen, 2013a, b) also attributed a prominent role in the actual functioning of organizations to the *parallel hierarchy of interpersonal influence* (leadership). According to Graen, each direct report in a hierarchy develops a unique strategic alliance (USA) with those in his/her collaborative network, which includes factors beyond the employment contract (Rousseau, 1995). Also, each collaborative network is linked to a more powerful one up hierarchy. In this way, organizations may achieve greater flexibility using different combinations of hierarchies. The prescription of LMX theory for building better organizations depends on proper designing of their *two interacting network hierarchies of authority and influence* (Graen & Schiemann, 2013).

Although many different theories of building a better organization are offered presently, this new one was supported by the success of Google's innovative designs to maximize the innovative performance of application teams (Bock, 2011). Google by establishing the conditions for team discovery and development has demonstrated the power of the two hierarchies these designs create in business units with protective, supportive and productive "cocoons." These cocoons have been shown to produce innovative products with fully engaged employees. LMX theory recognizes the secrets to Toyota, Walmart and Google's success and incorporates them in a model

TABLE 3.1 History of Network Alliance Making Thinking

1976	Leader–Member Exchange Framework	Graen
1977	Leader–Member Communications Networks	Schiemann
1977	Linking-Pin Quality: Leader–Leader–Member Exchange	Graen, Cashman, Ginsburg & Schiemann
1982	LMX Management Training	Graen, Novak, & Sommerkamp
1985	LMX Levels of Analysis: LMX Over TMX	Ferris
1988	Transformational Leadership	Avolio & Bass
1989	Charismatic Leader	Conger
1993	LMX Discourse	Fairhurst
1994	Cross-Cultural LMX	Graen & Wakabayashi
1995	First Revision of LMX	Graen & Uhl-Bien
1999	LMX Creativity	Tierney, Farmer & Graen
2003	LMX Alliance and Organizational Citizen Behavior	Hackett, Farh, & Song
2003	Reciprocal LMX Process: Fair Exchange	Uhl-Bien & Maslyn
2005	LMX Network	Sparrowe & Liden
2006	LMX Managerial Career Progress	Graen, Dharwadkar, Grewal, & Wakabayashi
2010	LMX Idiosyncratic Deals	Anand, Vidyarthi, Liden, C., & Rousseau
2010	LMX Leadership Identity	DeRue & Ashford
2010	Group Performance: LMX-Team	Naidoo, Scherbaum, Goldstein, & Graen
2012	LMX Meta-analysis Outcomes	Dulebohn, Bommer, Liden, Brouer, & Ferris
2013	LMX Leader–Leader Exchange	Liu, Tangirala, Ramanuja
2013	Back To The Drawing Board	van Knippenberg, & Sitkin
2014	LMX Squire Role	Weber & Moore

Note: Adapted from Graen, 2013b.

of collaborative-design mindset (Graen & Schiemann, 2013). Table 3.1 highlights key events in the development of the thinking (Graen 2013b).

In support of the theory, a large survey of employees of the U.S. Department of Veterans Administration found that the use of the LMX measure of existing alliance quality was a "silver bullet" in terms of employee reactions (Inis 2013). Those employees with a positive alliance with their manager complained significantly less about over 100 aspects of their employment conditions compared to those without.

STRATEGIC DESIGN TOOLS

In addition to the design breakthrough in collaborative overlapping alliance networks, our understanding of strategic competence has been superseded by the latest advancements in design thinking (Martin, 2004). No longer will a straightforward strength, weakness, opportunities and threats (SWOT) approach work satisfactorily. Today's Design School approach at Stanford demands that we stop solving the wrong problem by answering the questions of *why, how,* and *what* in that order (Breen, 2005). Before one defines a problem and proceeds to ideate, define prototypes, and test, one should empathize with those immersed in the problem situation. Simple SWOT thinking too often results in solving the wrong problems and suffering the consequences.

A critical step in design strategy is to empathize with the employees and customers. Asking *how might the employee or customer* is a reverse of what people are taught culturally and in school (Basadur, 2014). Designing strategy by executives involves competence in empathizing (walk a mile in the customer's shoes) to imagine the possibilities for meeting the needs and wants of customers in different situations. This design thinking is different from natural science thinking in that natural science seeks to find *universal laws of nature* in our environment. Design science thinking seeks *situation specific solutions* about improving our man-made local systems for coping with Mother Nature (March & Simon, 1958). For example, *How can we imagine a better way to serve our retail customers in small towns in the heartlands?* This was Sam Walton's design question. Sam also asked many design questions about his little, but growing retail business. *How can I design a better supply chain? How can I better serve my employees? How can I manage cash flow when my experts ask for the latest technology? What is a better growth model? What company songs should we sing?* (Walton & Huey, 1992).

A breakthrough in the design attitude was edited ten years ago by Boland and Collopy entitled *Managing as Designing* (2004). This book contains technical applications of design strategy by leaders in the field. Fortunately, a great deal of progress has been made in the last ten years. Training in design thinking is available through Stanford University, MIT and other leading schools offering design courses (Martin, 2004). Alongside these courses are decision-science courses that teach new methods in "competitive analytics" (Davenport & Harris, 2007) and "visual analytics" (Thomas & Cook, 2006) to gain competitive advantage. Analytical means "the extensive use of data, statistical and quantitative analysis, explanatory and predictive models to drive decisions and actions" (Davenport & Harris, 2007, p.12). These tools are used to stay ahead of the competition and also to find new markets (Grace, 2009).

Overall, this approach is recommended for research in executive development, strategy, problem solving, and more generally Organizational Behavior, Industrial and Organizational Psychology and Design Engineering. Moreover, case study methods generally could benefit from employment of these tools. Next, we consider the development of collaborative-design hierarchies.

BUTTERFLY DESIGN FOR MILLENNIALS

The landmark successes of the two cases outlined above may be informative and provide a context for the emerging "butterfly" designed organizations. One might see a problem in the successes of design thinking as too focused on whole situations, but it has produced protocols that work well across cultures as will be described below. Note that principles of "design thinking" are being incubated in university settings like the "Design School" at Stanford University (Breen, 2005). At the Design School, people from startups and large companies go to learn design thinking. Slowly, business is beginning to realize that this discipline of design is a useful approach to wealth creation and employing our talented children (Graen & Graen, 2009).

I recommend a new concept in organizational architecture called "collaborative-design mindset." It is based on providing a workplace that is attractive, friendly and understanding of what it takes to engage and retain a team's star performers. Based on the recent huge successes of Google with our youngsters, I predict that many successful firms will emerge with similar designs giving the needed edge to our stars team as they emerge from newly designed executive development pipelines. One of the key radical assumptions of the butterfly is that *employees and customers are the most important assets of a firm.* Also, be advised of the new terms used herein. For example, executive development work teams are housed in custom designed "cocoons" managed by trained "cocoons guardians" who take care of the security, harmony and resource support of all-star teams. One key characteristic of each cocoon is that all team members are required to enact a unique strategic alliance (USA) with at least each other including their "cocoon guardian" (manager). More will be described about these concepts, but know that the so-called "butterfly" designed firm is for a human-made "clan" of cocoons filled with only "siblings." Equally, important, cultural and ethnic differences are included in each unique strategic alliance.

New Employees

Eighty million young special talented professionals have emerged in the last 10 years and like butterflies they are ready to spread their life giving pollen from their school cocoons as they join the workforce. This should be a wakeup call. They have grown into something wonderful due to the pervasive possibilities of the newly developed miracle called the "the big bang" of the global Internet (Graen & Schiemann, 2013). Our millennial generations (1980s- present) have been incubated in a post-modern reality that differs from that of their parents. From their vantage in their childhood incubators, they experienced their world as "rich kids" (Stein, 2013). Their needs and wants were gratified almost immediately by their family and friends compared to their parent's. They were treated as "special" by almost all generation authority figures including the extended family, village, school, church, mall, recreation, health, and government. Child rearing practices, schools and even colleges were changed to be friendly and supportive of the special kids. As expected, our kids developed flattering self-concepts, global Internet competence and a strong service orientation (Graen & Schiemann, 2013).

These childhood incubators were close to B. F. Skinner's book entitled *Waldon Two*, in which only positive reinforcement was permitted. Although Skinner largely ignored human emotion in his theory, emotions played their role. Our children have been shielded from most suffering of everyday life by their families (Thomson, 2009). Overall, they offer tremendous potential for innovation. They are comfortable with multitasking, instant portable Internet searching, and adopting many new Internet applications. Their attention span is shorter and their rhythm has a faster beat than their parents. They are comfortable with authority figures until they are disrespected. They expect everyone to respect them as adults and any violation of egalitarian rule calls for an emotional reaction. Though we redesigned our child rearing practices, our schooling, our recreational organizations and our colleges using "helicopter parents," we have yet to redesign our workplaces. Thus, our mission is to design alternative prototypes of work environments that successfully attract, engage, enthuse, protect, support and retain both the talented star performers and bench players.

Fortunately, we are blessed with what we call *imagination* and *sentience*. These allow us to stretch our thinking beyond the present to alternative futures. Our so-called "butterfly design" may allow new professionals to forgo unnecessary suffering and improve the quality of their lives at work. The history of man is a story of great imaginations that improved our man-made world. As I stated above, we have advanced by discovering a primary design law, namely, *if something exists in our world, it exists in some quantity and can be measured*. With this law, we are able to discover new ways to measure our imagined future and use a collaborative-design mindset to produce successful new prototypes of workplaces which have improved our quality of life (Inis, 2013).

As Min Basadur counsels:

> Driving innovation through design thinking, 21st century manager's can accelerate their careers, "stand out among the rest," and differentiate themselves from their peers, most of whom are equipped with same basic creden-

tials and work in similar jobs in the organization or industry. They can also simplify coping with the bewildering pace of change in which they find themselves immersed by incorporating design thinking into their daily lives. First, they need to understand that design thinking is for everyone, not exclusive to people called "designers. (page TBD)

In one study, fifty millennial managers (1980-2000) in power industry were interviewed about what arrangements of the millennial's workplaces seemed significant to their performance (Thielfodt, 2012). She found that overall, the managers found that their generation loves teamwork and is willing to spend time building relationships. They easily collaborate, network and work with others. They prefer a more participative version of leadership and become frustrated with top down direction. These managers seek to be team players and focus on harnessing technology, adaptability and resourcefulness to contribute to team success. They also seek mentoring from those that they respect and trust, and value an ongoing learning environment for their teams. They are willing to take calculated risks and show a bias for action. Finally, they expect immediate gratification for their team efforts. Their performance rewards are similar to those of their parents, namely, (1) challenging, stimulating, meaningful work (59%), (2) growth (52%), (3) friendly workplace (49%), (4) pay (43%) and (5) balance (41%).

Laszlo Bock, Google's vice president of "people operations" states that his team discovered that a unit manager's behavior had much greater influence on employees' performance, retention and how they felt about their jobs than any other factor (Bock, 2011). Google's training of team managers paid off quickly by showing a 75% quality improvement of the worst performing managers. The findings that many good things happen when managers practice proper leadership with those who work for them, has been a consistent research result in studies of LMX (Graen, 2013a, b; Graen, 2014).

Google's "People Operations" department should not be seen as the old Human Resource Department, but rather as a new function. People operations are more similar to the other functional departments such as finance, purchasing, production, and sales in being based on data mining. This department has the responsibility for using data mining analytics to train executives regarding the costs and benefits of various decisions. Rather than being overwhelmed by "war stories" and "gut reactions," properly analyzed and interpreted data are presented as a base for informed decisions. Data are often collected and available before a problem emerges. This anticipation of the dynamics of an innovative company produces unexpected competitive advantages. Studies of managerial leadership consistently reveal the finding that professional associates, from many different firms, live and work in relatively more productive, engaging, and enjoyable environments (developmental cocoons) when they have negotiated a unique strategic alliance

(USA) with their managers compared to those who have no such alliance (Dulebohn, Bommer, Liden, Brouer, & Ferris, 2012). The differences between the world of work for those who possess this alliance and those who do not extend to almost all characteristics of the work itself and its surroundings. Millennial star performers are likely to be attracted to executive development pipelines that incorporate the butterfly designs. Google employed design science methodology to demonstrate the power of strategic design integrated with network-alliance making and has continued to exploit it for the welfare of its customers, employees, managers and shareholders.

According to Bock (2011), his "People Operations Program" (POP) functions are a component of the strategic core of Google[1]. Moreover, this component describes the "conditions necessary" to develop "creative, committed, and innovative teams." This component has been called many things from charismatic leadership to company culture. I call it "butterfly design" because it describes a physically and psychologically safe and secure culture for business unit cocoons, which promotes the full display of the free play of the sights, sounds, touches, smells and creative juices within the human-made world. Without these "cocoons," the potential "*what ifs*" are not asked due to anxiety avoiding reactions. These creative cocoons include the signals of (a) a selected engaging mission that rises from the excitement of Googlers, (b) an enabling process that is open and honest regarding almost everything, and (c) robust, data-driven transparent decision making across pay grades and cocoons. Millennials at Google are delighted to work in such innovation studios. But it still needs work.

Specifications for a Butterfly Organization

The five butterfly design characteristics found in Steiber's year-long residential study inside Google were (Steiber, 2011):

1. An innovative and flexible culture and management system that replaced structural rules with heuristics and command and control with peer-like negotiating latitude between associates across pay levels.
2. A company which valued employees and customers as the most important assets by selecting the best, treating the best as main contributors who must be given proper career opportunities, rewards, and trusted with inside information at all times.
3. Bosses at all levels were encouraged to work with their people in manners that were appropriate to their uniqueness by tailoring their mentoring, empowering coaching, and clearing away various impediments.
4. Balancing the emphasis of innovations and operational excellence by fostering the development of subcultures equally valued.

5. A fundamental commitment to searching for externally developed technical innovations, cooperate alliances with leading universities/researchers, and investing in new technologies and ventures. (Adapted by author from Steiber, 2011)

Google's collaborative design approach to the issue of "ambidexterity" in firm strategy (O'Reilly & Tushman, 2013) clears away much of the strategic ambiguity. Their approach does not seek a universal strategic mix of "explore" and "exploit." Rather, it allows for different design mixes appropriate to the actors, behaviors and context (ABCs) of the firm (March, 1991). The proper mix must be found at critical times, because wonderful exploitation of mature products without the development of new products may signal the end of the firm. O'Reilly and Tushman (2013) present many examples of companies attempting to do both as their mature products have declining demand. For reasons currently not understood, some firms survive with this ambidextrous strategy and others do not. The point here is that different combinations of the ABCs may have different relations to sustainability and other measures of firm health. Design thinking renders this possibility moot by tailoring a design for a particular firm. In this case, one size is not designed to fit all firms.

Google successfully built leadership cocoons for its teams by asking a fundamental design thinking question, namely, "*What if*" we designed cocoons that provide the motivational environment to attract the best technical talent and enhance their collaborative teamwork. Using the methods of design, Google discovered the boundary conditions of their company culture. These included Hackman's (2011) team architectural structure, Graen's (2013a b, Graen 2012) team dynamics, and Shuffler, Burke, Kramer, & Salas, (2013) team enhancements. In short, the recipe included continually adjusting conditions to provide what talented associates seek most in their productive lives. Not surprisingly, Google found that the top conditions were *trust, psychological safety and opportunity, openness and honesty*, and *freedom to demand data-based answers*. The boundary conditions have been confirmed by recent research (Graen, 2014), but only Google displayed the courage to immerse itself in the knowledge-driven organization form. Google's innovative methods invite their use of other functions such as management of customer, supplier, sales, and the top management team.

John Sullivan (2013) lists his "top 10" people management practices of Google however, he fails to provide the theory behind these practices. Researching the appropriate research literature in management reveals the whys and wherefores of Google's innovative practices. Relevant tested theories need to be described for each of the innovative practices. The new prescription for Google managers' team performance and retention was "one-on-one" coaching with trust, interest and frequent "personalized" feedback first and foremost and superior technical knowledge a distant second place. To track this, managers are rated every 6 months by those who work for them. One theory was supported by these "project oxygen" team research results (Graen, 2014).

Google's company culture appears to question the validity of the popular stereotype of highly technical people as being characterized as discounting human emotions as irrational (Dr. Spock of Star Trek). Google's culture is based on the assumption that highly technical people can be more productive and innovative in more emotionally safe, supportive and collaborative cocoons managed by trusted, attentive, and enabling coaches with personalized business alliances. Moreover, the conditions promoting different "*What ifs*" benefit from designing different cocoons for each.

Team members should learn about each other's strengths and weaknesses, promises and threats and proactively interact to render interactions purposeful and instructive. Additional processes need to be reviewed for hiring, training, placing in cocoons with data-based conditions including team development and leadership programs, and maintaining productive environments of cocoons against disruptions. The organization design depends on the product or service of interest. For the invention and development of new apps, a Google-like design would be appropriate. Even better would be a design that leap-frogs Google. For breakthrough research, our external research alliance system combined with cocoons would be a good place to start. For manufacturing and incremental innovations, the Toyota R&D system would be a proper beginning. Finally, for maximum collaboration, Walmart and Crossmark are leadership hard to beat.

An additional proposed characteristic of the butterfly cocoon is called the "manager's gorilla support." This support of cocoon residents is provided by a manager when permitting a direct-report the informal acquisition of resources discouraged by the hierarchy organization. *The Soul of a New Machine* was a great example of this action (Kidder, 2000). This is an expected part of a cocoon manager's risks when critical resources are needed. Critics often mistakenly make the attribution that gorilla support is "political" (Ferris & Treadway, 2012). In fact, gorilla support is one of the factors that keeps the formal organization flexible and alive. Of course, after it is exercised, it must be justified up the hierarchy and must be used wisely.

TRADITIONAL APPROACHES

Finally, a review of progress of traditional approaches in establishing research-based prescriptions for successful design of better management development organizations has produced a plethora of speculative theories (Zaccaro, 2001). Research over 100 years has produced long lists of attributes categorized as *cognitive, social,* and *self-motivational* requirements and corresponding executive attributes (Zaccaro, LaPort, & Irwin, 2013). *Cognitive* requirements are further divided into *direction setting and planning* and *internal operational management,* each with multiple subcategories. The corresponding executive attributes numbered 15 for *cognitive,* 12 for *social,* and 8 for *self-motivation.* Unfortunately, the sum result of many meta-analyses and reviews found *no universal laws* of executive success. Explanations abound for these results (Mumford, Zaccaro, Harding, Jacobs, & Fleishman, 2000; DeRue, Nahrgang, Wellman & Humphrey, 2011). Several alternatives have been suggested to find something more useful such as, configural modeling and process modeling (Zaccaro, 2007), larger sets of attributes (Zaccaro, et al., 2013), multi-level studies (Zaccaro, 2001), shared responsibility (Pearce & Conger, 2003) star performers (Aguinis & O'Boyle, 2014) and new theories (Zaccaro, et al., 2013). Day (2013) agreed that the field remains at a primitive stage of development and hopes for great progress by 2050.

An alternative taxonomy of approaches using traditional research methods might consist of (1) a*ttributes of successful executives;* (2) *characteristics of successful executive team leaders;* (3) *contexts of successful executive teams.* Under (1) *executive-centric:* Zaccaro (2001), Hogan and Judge (2013), Borg, (2012), Day (2000). Under (2) *executive team-centric:* Hollander (2009), Salas, Wilson-Donnally, Sims, Burke and Priest, (2007), Graen, (2013b). Under (3) *context-centric:* Klimoski (2013), Graen & Graen (2013). In contrast, the collaborative-design mindset approach would concentrate on combined (2) and (3) above to produce (4) *executive teaming in context:* Boland and Collopy, (2004), Grace and Graen (2014), Martin (2004). In sum, the argument is that a design science approach integrated with network alliance making supplies a new kind of thinking and new tools for managers. The traditional approach has failed. Hence, a new path needs to be considered.

This assumes that a better organization is characterized as competent and able to contribute to affective strategies for the company and execute strategies using design protocol and networks of alliances. This also assumes that each company has its unique customers, employees, suppliers, distributors, shareholders, and culture (Schneider & Babera, 2014). Design methodology uses data on each of these and in combination to identify developmental needs of a particular firm. All this should be part of formal strategy.

CONCLUSION

The Center for Creative Leadership (CCL) surveyed the best practice in building a better organization (McCauley, 2008) from three perspectives: the specification of developed executives, developmental methods, and fostering practices. This large survey produced more refined questions:

1. Are there a few core capabilities that equip organizations to be more effective in particular contexts?
2. Do certain organizational designs best develop each capability?
3. Is the world of work changing?
4. Do these changes have impact? (adapted from McCauley, 2008)

My reply to these questions based on our project presented in this chapter, are: (1) yes, (2) yes, (3) yes, (4) yes. I predict that our organizations will become more effective due to new designs using a "collaborative-design mindset" (strategic design thinking integrated with network alliance making). Our better executives will be educated in college and trained at the firm using a variety of learning experiences discussed herein including team and peer collaboration (Grace & Graen, 2014). He/she must be prepared to adapt successfully to the rapidly developing set of discontinuous changes from all directions including generational and the cultural. Our man-made world continues to accelerate its speed of change. Our best hope rests on our ability to innovate and keep pace. We must continue to search for the laws of nature and for what works for us here and now. Our innovation edge and the promise of new generations are terrible things to waste.

Progress of our traditional (banker) thinking approach to building a better universal organization have been disappointing compared to that of collaborative-design mindset (Graen, 2014). Note that recently Google bought Nest Labs for $3.2 billion. The rationale given was that Google has the financial resources and functions like human operations, customer service and law. These internal support networks were seen as enabling Nest to grow and expand globally. It also was seen as conforming to Google's vision of a connected world that goes from pocket machines to transportation devices to homes (Winkler & Wakabayashi, 2014). Nest was the 20th purchased by Google in the past year (Stone, 2014). This is one example of the power of the butterfly design.

A key difference between the two approaches is that traditional ones focus on discovering core capabilities that are effective in many contexts (universal), but the butterfly approach seeks the capabilities that are effective for a particular context. This context includes parameters for *a*ctors, *b*ehaviors and *c*ontext (ABCs). Even Toyota managers, who are described as generalists, are required also to be specialists in particular areas. The same is the case for Walmart and Google. The design question becomes: For what functions do you want to build a better organization? Please be specific. When the target functions are incompatible, perhaps two different organizations will be required. It seems a bit romantic to seek the charismatic organization that is effective in all contexts. I admit that I can organize my research team, but I struggled to manage my university department.

Finally, specifications for an employee friendly and innovation producing organization has been adapted from Steiber (2011) to design the butterfly organization. The butterfly model was selected to connote that lasting innovations in the design of organizations of the future depends on multiple possible meaningful attributes for different cultural groups of designers, producers, distributers, customers, regulators, communities and bankers throughout the supply chains (Hurst, 2012). (Recall that the monarch butterfly requires four different generations to endure one year.) This theory stresses the functions and structures required to construct the flexible and adaptive organization that can change direction as required. In sum, for the future, design thinking seems more promising than traditional thinking. My word to the wise is please plan to leap-frog the competition by learning about the new and wonderful collaborative-design mindset. Along with Robert Austin, I support the new "Spring of the Millennials." Never give up.

NOTE

1. Both Toyota and Walmart are designed in this manner.

REFERENCES

Aguinis, H., & O'Boyle, E. Jr. (2014). Star performers in the twenty-first century organizations. *Personnel Psychology.*

Anand, S. , Vidyarthi, P. R., Liden, R. C., & Rousseau, D. M. (2010). Good citizens in poor quality relationships: Idiosyncratic deals as substitutes for relationship quality. *Academy Management Journal, 53*(5), 970–988.

Avolio, B. J., & Bass, B. M. 1988. Transformational leadership, charisma and beyond. In J. G. Hunt, B. R., Baliga, H. P., Dachler, & C. A. Schriesheim (Eds.), *Emerging leadership vistas* (pp. 29–49). Lexington, MA: Heath.

Barnard, C. I. (1938). *Functions of the* Executive. Cambridge: Harvard University Press.

Basadur, M. (2014). Taking the mystery out of design thinking. In M. Grace & G. Graen (Eds.), *Millennial management: Designing the future of organizations. LMX leadership: The series* Vol IX. Charlotte, NC: Information Age Publishing.

Bock, L. (2011). Passion, not perks. *Think-Insights-Google. Think Quarterly: The People Issue.* Google Incorporated.

Boland, B. J., & Collopy, F. (2004). *Managing as designing,* Stanford University Press.

Borg, C. (2012). *unique research inner life of Google.* Stockhelm: Eurekalert Publishing.

Breen, B. (2005, April). The business of design. *Fast Company.* Retrieved December 26, 2008 from http://www.fastcompany.com/magazine/93/design.html.

Conger, J. A. (1989). *The charismatic leader.* San Francisco: Jossey Bass.

Davenport, T. H., & Harris, J. G. (2007). *Competing on analytics.* Boston: Harvard Business School.

Day, D. (2000). Leadership development: A review in context. *The Leadership Quarterly, 11*, 581–613.

Day, D. (2013). Training and developing leaders: Theory and research. In M. Rumsey (Ed.), *Oxford handbook of leadership* (pp. 76–93). Oxford, UK: Oxford University Press.

DeRue, S. P., & Ashford, S. J. (2010). Who will lead and who will follow? *Academy of Management Review, 34*(4), 627–648.

DeRue, D. S., Nahrgang, J. D., Wellman, N., & Humphrey, S. E. (2011). Trait and behavioral theories of leadership: An integration and meta-analytic test of their validity. *Personnel Psychology 64*, 7–52.

Dulebohn, J. H., Bommer, W. H., Liden, R. C., Brouer, R., & Ferris, G. R. (2012). A meta-analysis of the antecedents and consequences of leader-member exchange: Integrating the past with an eye toward the future. *Journal of Management, 38*, 1715–1759.

Fairhurst, G. T. (1993). The leader-member exchange patterns of women leaders in industry: A discourse analysis. *Communication Monographs, 60*, 322–351.

Ferris, G. R. (1985). Role of leadership in employee withdrawal process: A constructive replication of Graen's study. *Journal of Applied Psychology, 70*, 777–781.

Ferris, G. R., & Treadway, D. C. (2012). *Politics in organizations: Theory and research considerations*, New York: Routedge/Taylor and Francis.

Grace, M. (2009). Development of design project teams and their supporting resource networks for the knowledge era. In G. Graen & J. Graen (Eds.), *Predator's game-changing designs: Research-based tools. LMX leadership: The series* Vol VII, (pp.1–18). Charlotte, NC: Information Age Publishing.

Grace, M., & Graen, G. B. (2014). What if we designed A MBA for the future? *Decision Science.*

Graen, G. B. (1976). Role making processes within complex organizations. In M. D. Dunnette (Ed.), *Handbook of industrial and organizational psychology*, (pp. 1201–1245). Chicago: Rand-McNally.

Graen, G. B. (2012, September). Chinese executive leadership teams need both stars and support players. *Harvard Business Review, China.*

Graen, G. B. (2013a). Overview of future research directions for team leadership. In M. G. Rumsey (Ed.), *The Oxford handbook of leadership* (pp. 167–183). Oxford, UK: Oxford University Press.

Graen, G. B. (2013b). The missing link in network dynamics. *The Oxford handbook of leadership* (pp. 359–375). M. Rumsey (Ed.). Oxford, UK: Oxford University Press.

Graen, G. B. (2014). What have we learned that is intersubjectively testable regarding the leadership process and leadership-performance relations? *Industrial and Organizational Psychology Perspective on Science and Practice.*

Graen, G., Cashman, J., Ginsburg, S., & Schiemann, W. (1977). Effects of linking-pin quality on the quality of working life of lower participants. *Administrative Science Quarterly, 22*, 491–504.

Graen, G. B., Dharwadkar, R. Grewal, R., & Wakabayashi, M. (2006). Japanese career progress over the long haul: An empirical examination. *Journal of International Business Studies, 37*, 148–161.

Graen, G. B., & Graen, J. A. (2009). *LMX Leadership: The Series*. Predator's game-changing designs: Research-based strategies. Vol. 7. Charlotte, NC: Information Age Publishing.

Graen, G. B., & Graen, J. A. (2013). *LMX Leadership: The Series*. M*anagement of team leadership in extreme context: Defending our homeland, protecting our first responders. Vol. VIII.* Charlotte, NC: *Information Age Publishing.*

Graen, G., Novak, M., & Sommerkamp, P. (1982). The effects of leader-member exchange and job design on productivity and satisfaction: Testing a dual attachment model. *Organizational Behavior and Human Performance, 30,* 109–131.

Graen, G. B., & Schiemann, W. (2013). Leadership-Motivated Excellence Theory: An Extension of LMX. *Journal of Managerial Psychology, 28*(5), 452–469.

Graen, G. B., & Uhl-Bien, M. (1995). Development of leader-member exchange (LMX) theory of leadership over 25 years: Applying a multi-level multi-domain perspective. *Leadership Quarterly, 6*(2), 219–247.

Graen, G. B., & Wakabayashi, M. (1994). Cross-cultural leadership-making: Bridging American and Japanese diversity for team advantage. In H. C. Triandis, M. D. Dunnette, & L. M. Hough (Eds.), *Handbook of industrial and organizational psychology* (Vol. 4, pp. 415–446). New York: Consulting Psychologist Press.

Hackett, R. D., Farh, J. L., & Song, L. J., (2003). LMX and Organizational Citizenship Behaviour: Examining the links within and across Western and Chinese Samples. In George B. Graen (Ed.), *Dealing with diversity, LMX Leadership: the Series* (Vol. 1, pp. 219–264). Information Age Publishing, Greenwich, CT.

Hackman, J. R. (2011). *Collaborative intelligence: Using teams to social hard problems.* Berrett-Koehler, San Francisco, CA.

Hackman, J. R., & Oldham, G. R. (1976). Motivation through the design of work: Test of a theory. *Organizational Behavior and Human Performance, 16,* 250–279.

Hage, J. (2011). *Restoring the innovative edge: Driving the evolution of science 2011 and technology.* Stanford, CA: Stanford Business Book.

Hogan, R., & Judge, T. (2013). Personality and leadership. In M. Rumsey (Ed.), *Oxford handbook of leadership* (pp. 7–46). Oxford, UK: Oxford University Press.

Hollander, E. P. (2009). *Inclusive leadership: The essential leader-follower relationship.* New York: Routledge.

Hui, C., & Graen, G. B. (1997). Sino-American ventures in mainland China. *Leadership Quarterly, 4,* 451–465.

Hurst, N. (2012, May). Big companies buying design firms. *Industry Week.*

Inis, S. (2013). *Results of USD of Veterans administration employee survey.* E-mail to Graen, March.

Isaacson, I. (2011). *Steve Jobs.* New York, NY: Simon and Schuster.

Kidder, T. (2000). *Soul of a new machine.* San Francisco: Back Bay Books.

Klimoski, R. (2013). When it comes to leadership, context matters. In M. Rumsey (Ed.) *Oxford handbook of leadership* (pp. 267–291). Oxford, UK: Oxford University Press.

Liu, W., Tangirala, S., & Ramanuja, M. (2013). The relational antecedents of voice targeted at different leaders. *Journal of Applied Psychology, 98*(5), 841–851.

March, J. (1991). Exploration and exploitation in organizational learning. *Organizational Science 2,* 71–87.

March, J. G., & Simon, H. A. (1958). *Organization.* New York: Wiley.

Martin, R. (2004). The design of business. *Rotman Management, 5*(1), 6–10.

McCauley, C. D. (2008). *Leadership development: A review.* Greensboro, NC: Center for Creative Leadership.

Merton, R. K. (1957). *Social theory and social structure.* Glencoe, IL Free Press.

Mumford, M. D., Zaccaro, S. J., Harding, F. D., Jacobs, O. T., & Fleishman, E. A. (2000). Leadership skills for a changing world: Solving complex social problems. The *Leadership Quarterly, 11*, 11–35.

Naidoo, L. J., Scherbaum, C. A., Goldstein, H. W., & Graen, G. B. (2010). A longitudinal examination of the effects of LMX, ability, and differentiation on team performance. *Journal of Business and Psychology, 26*(3), 347–357.

NLRA. (1935). Title 29, Chapter 7, Subchapter II, US Code, Federal law US, *Library of Congress.*

O'Reilly, C. A., & Tushman, M. L. (2013). Organizational ambidexterily: Past, present and future. *Academy of Management Perspectives. 27*(4), 324–338.

Pearce, C. L., & Conger, J. A. (2003). All those years ago: The historical underpinnings of shared leadership. In C. L. Pearce & J. A. Conger (Eds.), *Shared leadership: Reframing the hows and whys of leadership* (pp. 1–18). Thousand Oaks, CA: Sage.

Pfeffer, J. (2013). You're still the same: Why theories of power hold over time and across contexts. *Academy of Management Perspectives, 27*(4), 269–280.

PWC. (2013). PWC's next Gen: A global generational study. www.pwc.com.

Rousseau, D. (1995). *Psychological contracts in organizations: Understanding written and unwritten agreements.* Thousand Oaks, CA: Sage Publications.

Salas, E., Wilson-Donnelly, K. A., Sims, D. E., Burke, C. S., & Priest, H. A. (2007). Teamwork training. In P. Carayon (Ed.), *Handbook of human factors* (pp. 803–822). Mahwah, NJ: Erlbaum.

Schiemann, W. A. (1977). *Structural and interpersonal effects on patterns of managerial communications: A longitudinal investigation.* S. Rains Wallace Award Doctoral Dissertation, University of Illinois.

Schneider, B., & Barbera, K. M. (Eds.) (2014). *Oxford handbook of organizational climate and culture.* New York: Oxford University Press.

Shuffler, M. L., Burke, C. S., Kramer, W. S., & Salas, E. (2013). Leading Teams: Past, Present, and Future Perspectives. In M. G. Rumsey (Ed.), *The Oxford handbook of leadership* (pp. 144–166). London, UK: Oxford University Press.

Sparrowe, R. T., & Liden, R. C. (2005). Two routes to influence: Integrating leader-member exchange and network perspectives. *Administrative Science Quarterly, 50*(4).

Steiber, A. (2011). Society for human research management. *Research report on Managerial Leadership Needs.* New York: SHRM.

Stein, J. (2013). The new greatest generation: Why millennials will save us all. *Time.* May 20.

Stone, B. (2014). The social network for startups. *Bloomberg Businessweek.* January 20.

Sullivan, J. (2013). How frugal is using people analytics to completely reinvent HR. *TLNT.com.*

Thielfodt, (2012). Future leadership: Looking at todays millennial managers and how they lead. *TED.*

Thomas, J. J., & Cook, K. A. (2006). A visual analytics agenda. *IEEE Transactions on Computer Graphics and Applications, 26*(1), 12–19.

Thomson, L. J. (2009). Review of the trophy kids grow up by Ron Alsop, *Academy of Management Learning and Education, 8*(3), 464–466.

Tierney, P., Farmer, S., & Graen, G. B. (1999). An examination of leadership and employee creativity: The relevance of traits and relationships, *Personnel Psychology, 52*(3), 591–620.

Uhl-Bien, M., & Maslyn, J. (2003). Reciprocity in manager-subordinate relationships: Components, configurations, and outcomes. *Journal of Management, 29*(4), 511–532.

van Knippenberg, D., & Sitkin, S. B. (2013). A critical assessment of charismatic–transformational leadership research: Back to the drawing board? *Academy of Management Annals, 7*(1), 1–60.

Walton, S., & Huey, J. (1992). *Sam Walton: Made in America.* New York, NY: Doubleday.

Weber, J. M., & Moore, C. (2013). Squires: Key followers and social facilitation of charismatic leadership. *Organizational Psychology Review, 4,* 99–123.

Winkler, R., & Wakabayashi, D. (2014). Google pays $3.1 billion for Nest Labs. *Wall Street Journal,* January 14.

Zaccaro, S. J. (2001). *The nature of executive leadership: A conceptual and empirical analysis of success.* Washington, DC: American Psychological Association.

Zaccaro, S. J. (2007). Traits based perceptions in leadership. *American Psychologist, 62,* 6–16.

Zaccaro, S. J., LaPort, K., & Irwin, J. (2013). The attributes of successful leaders: A performance requirement approach. In M. Rumsey (Ed.). *Oxford handbook of leadership* (pp. 11–36). Oxford, UK: Oxford University Press.

CHAPTER 4

(RE)STARTING FROM SCRATCH

Andrea Cifor
Sarah Chana Mocke
Microsoft

ABSTRACT

Technological innovation has driven our culture, the types of work that we do, and the employment practices of businesses for centuries. Who could have imagined the incredible numbers of authors self-publishing books and earning a respectable living in the days that preceded Gutenberg's printing press? In those times reading was something the elite in society might do, and writing was an art form left to a very few specialists that concentrated their efforts on religious texts. Even as the general public dismissed portable books, and the ruling elite and clergy attempted to demonize the printing press with allegations such as the pen being comparable to a virgin and the printing press a whore, an invisible movement was underway. Today it is beyond imagination to consider that people might not need to read, write and communicate through email, instant messaging, social networks and the Internet. The incredible democratization of knowledge and tools has created a level playing that is starting to erode the foundations of traditional business and their employment practices. Entirely new careers have been established as a result. Careers we might never have been able imagine just a few decades ago.

Millennial Spring: Designing the Future of Organizations, pages 65–83
Copyright © 2014 by Information Age Publishing
All rights of reproduction in any form reserved.

This chapter describes some of the challenges that we face in the technology sector today as we work to continually adjust to a rapidly changing landscape. It describes the remarkable agility that we need to develop and introduces some of the social behaviors that we must depend on to be successful. The changes engulfing technology-related employment today are merely precursors for the changes that will take place in many other industries too. Whether you've spend decades in gainful employment and are feeling nervous about the future, or you're just entering the workforce today, we hope that this chapter will provide you with some direction and insight into the future world of employment.

INTRODUCTION

For those entering the IT industry in the late 1980s a new age of Ethernet, Microsoft Windows, Novell Netware and PC-based client/server computing started to take hold. Young upstarts, were walking through the hallowed, impenetrable halls of corporate IT threatening the older establishment narrative. Just as their predecessors replaced typewriters, calculators and filing systems; this generation moved beyond impersonal, unwieldy task-based mainframes and terminals to the personal computer that was struggling to gain ground as a serious tool beyond the menus, word-processors, spreadsheets, databases and occasional page layout programs. The corporate world became computer literate outside of an IT department and resumes abounded with lists of "packages."The 90s introduced a new breed of computer whiz kid that learned by doing. Startups gobbled up anyone that could jump behind a keyboard and assimilate into the "dot.com" boom. Everything we knew in a brick and mortar world reared its ugly head into a series of strange, flailing flops and failures as each group tried to come up with a unique idea that would sell. Those with established brands under their belts sat back and orchestrated the frenzy participating in bidding wars for the best developers, testers and IT staff. The old guard funded the new and often benefitted heavily while the dot.commers suffered the dot. bombs and were left like a group of rats fighting for a piece of cheese trying to find their next job.

As the wreckage cleared, creativity took a hiatus. Bigger companies sucked up the broken techies at dimes for dollars over previous salaries. Like a magic reboot, the industry abstained from the dot.com addiction and settled onto a healthy diet guaranteed to build more solid foundations. Gone were the mad days of invest in anything because it sounded cool. The predecessors brought discipline back into the picture and technologists were beholden to their rules.

Diving into the new millennium a slow trickle of devices enabled a new flexibility in and out of the office. While the laptop and cellular phone were common tools, prior to 2000 they were limited in their capabilities and clunky with low battery life and limited functionality. As we progressed through the 2000s and into the 2010s we've witnessed the boundaries of offices, meetings and buildings melt away to free the masses to work from just about anywhere imaginable.

IT departments, that were used to complete control, felt undermined and increasingly disempowered. Panic and uncertainty triggered fight or flight responses, which resulted in attempts to regain command and control of the departure of IT employees to more hierarchical environments. Some attempted to maintain their relevance by reframing old narratives and couching older philosophies. Phrases such as, "no one got fired for buying IBM," were repeated frequently and loudly almost as if it was a security blanket. It was a form of FUD (fear, uncertainty, and delight—a means to market by sewing doubt about the efficacy of solutions and products from competitors) that brought comfort to those who sought job security because it made them feel relevant and useful. The question that starting tumbling in the collective subconscious of the IT professional masses was whether job security itself was becoming as obsolete as yesterday's systems and "packages" due to the rapid pace of change and development in technology. Of course, many responded to the uncertainty of change by embracing it. The edge-of-the seat thrill ride, and the almost childlike awe of new technology, is a necessary part of the experience for many that would otherwise become bored. Job security is provided by the inherent instability of the IT industry, rather than convergence. The dichotomy between achieving consistent and durable IT outcomes with technology and the long-term viability of experienced IT architecture practitioners that are dependent both on their ability to adapt to uncertainty and instability and to deliver results regardless is fascinating.

The anticipation of what is next draws us in and engages us. Just as these awkward moments of the unknown are vehicles for driving an outcome, it is quite possible that they also are equally effective in driving creativity.

Is it possible that we die a kind of career death when we decide to stop learning and adapting to the World around us? That seems to be the moment that older generations decide to start handing over the reins. Some stick around, dying of a thousand cuts, as they first attempt to ignore the changing world, and then more steadfastly resist it. Perhaps this is how it has always been, but in latter times it seems that sweeping changes in careers are occurring with greater frequency necessitating career progressions that are far from the slow, incremental changes that many of us would undertake to stay relevant previously. Disruption is the term on everyone's lips. We're no longer expected to simply adjust to using new tools as part of our trade. The trade itself is different.

Technology industries have preceded dramatic changes in labor over the past two centuries. From mechanization to electrification, through to the ubiquity of computerization and the already widely present, but expected greater adoption of robotics and machine learning, the world is on the move.

In 1441, Gutenberg invented the printing press, and people were dismissive, contending that a 1286 page book, The Gutenberg Bible, would simply be too heavy for people to carry. It was generally accepted by the public, at large, that they did not need to read books and they were dismissive. Clergies and governments feared losing control of information. To them the pen was a virgin, and the printing press was a whore.

It is possible to draw many parallels between those times, and what is present with social networking today. The crackdown on information, the early adopters, mass adoption, the prophets of doom and those who attempt to control the information flow as much as possible all seem somewhat familiar. Naysayers will insist that everyone will forget how to talk to each other, that people will never visit one another, and that most of us might stop attending concerts and lectures altogether. Indeed, they'll insist that people will never need to leave home at all. It's amusing to recognize that many similar objections were present during the times of such inventions as the telephone, gramophone and email too!

Just as was the case with the effect of the printing press on the publishing industry, it is anticipated that entirely new careers will surface in technology industries too. As technology evolves, so does everything else. Yet, somehow, certain realities ring true; people will continue to find ways to earn a living, bringing shelter and security to their families and others.

The key to doing so will lie in embracing an ever-evolving career that is able to span disruption. In this chapter we'll attempt to articulate the problems that you might face in the IT sector, and it's reasonable to assert that these are similar to the kinds of changes people employed in many other industries might face.As the writers of this chapter, we hope that you will continue unaffected as you face the influx of the next generations entering the workforce. We hope that you will learn to embrace and adopt what they bring to the table and find ways to relate what you have learned from experience for the benefit of them and all those around you.

It's an edge of the seat thrill ride. Welcome to the front row!

THE LIGHT BEGINS TO FADE

The room was getting darker. At first it was imperceptible, but you knew the light would fade into darkness eventually. The warning signs were there. Sometimes they weren't just warning signs, they were obstacles that you didn't notice that caused you to trip and stumble. You thought stumbling wasn't so bad. It was good to have a reminder now and then that you had to keep your skills up to date. Young people enter the workforce every day armed with approaches, skills and tools that continually represented a threat to your relevance. Until now it hadn't really affected you much. Why worry now? Technology had always been the same. We use it to solve business problems. It provides some kind of intrinsic value or people don't buy it. For it to work properly skilled workers are required. You are one of those: a highly sought after, skilled and talented resource in short supply. You rationalize, and rationalize so much that ultimately you believe it. You become the proverbial ostrich with its head stuck in the sand. They are younger, they surround you now and look at you knowingly, and callously talk amongst themselves about you needing to "retire" or "move on."

Every decade or so a seismic shift occurs in the technology industry. It creates a chasm so wide that if you don't take the leap of faith required to succeed you will slowly fade away into oblivion. You saw this yourself, your first job. You were competitive and ripe and ready to take over the world from the old guard, and you did. You see very few of those old guys around today—you and your peers replaced most of them. Few of them were able to survive and flourish in the new world you created.You entered the IT industry at an early age, fresh from school and ready to change the world, just like the new class of fresh faces ready to replace you. You were mentored by COBOL programmers who were made obsolete by C#, Java, Ruby. The list goes on. Only a few of those archaic programmers survived, the few who did were heavily sought after and made their companies a fortune doing work no one else was capable of doing.This was not an experience unique to the USA. As Sarah Chana Mocke describes the beginning of her career,

As a 19-year old I gave my uncle that knowing look. I was an upstart, working for a distributor in South Africa that bet its whole business on selling Microsoft and being the best technically at delivering their solutions. The year was 1989. It wasn't just software we were selling. We were changing the world. A few short years later my uncle and I started engaging in a battle between two major operating system for computers, Microsoft Windows 95 and IBM OS/2 2.0. I was torn. I'd been engaged in a fight between two networking products, Novell Netware and Microsoft LAN Manager, for quite a while, and had been really successful deploying LAN Manager in competitive situations and winning. LAN Manager ran on a "real" (my opinion) operating system, Microsoft OS/2 1.31. When Microsoft shifted their strategy I had to make a hard choice, and I chose Microsoft. My uncle thought I was nuts, but I wanted to change the world. The open source movements and the Internet came along in time too, and they were just as scary. Again I signed up to change the world. It seemed so simple.

Similar to Sarah, you are faced with a seismic shift. You have to develop the skills and traits to embrace a new world in which deep technical knowledge is becoming commoditized or encapsulated in devices and services, and you need to do new things. Your shelter and security depend on it.All the commoditization and encapsulation has created a new generation of consumers that simply take technology for granted. If it doesn't work it gets returned. No one needs you to fix it. The next generation clambering at the door knows how to upgrade firmware on their devices without hard-to-fathom text-based software, boot floppies, cables and EPROM programming.As you look to the past decades of technology, you should pat yourself on the back, and be proud that this industry has reached a point that people are able to use it without requiring experts all around them constantly. Of course, that does pose a dilemma too. It creates uncertainty and deep anxiety about where your future, in fact anyone's future income will be derived. You can look at that challenge pessimistically, or you can work out what might be needed in the future.This dilemma is shared by each generation of technologists as a new generation leaves college and enters the work force as illustrated by Sarah,Part of my present job requires me to look forward in this way. Determining the skills someone may need 18 months to two years from now is no easy task in an industry sector that is so full of disruption. As much as I like to think that I know what the new world order for IT may look like, I simply can't know. Enormous companies seem to rise and fall in less than a decade, which makes it hard to bet one's career or a future on anything that seems stable. Rather, it seems, livelihoods can only be determined by finding ways to service needs regardless of the underlying technologies harnessed at any given time. That seems "fluffy" and intangible. How can someone be an expert at anything if you are not worrying about the details of what lies beneath the seemingly superficial veneer?

At first glance, the conclusion may seem simple. Servicing needs is pretty fundamental to most business endeavors. However, in the IT industry, this level of abstraction actually makes solving problems more complex. You and your peers in IT are faced with abstraction, which tends to depend on more and more moving parts that no longer fall within your span of control. Despite extensive efforts to standardize across the industry, IT vendors' must differentiate in order to attract customers. This usually results in their competitive advantages being encapsulated within their technologies and products.

In the past, enterprises would standardize as much as possible by purchasing technologies from a limited number of IT vendors. As an IT worker, you would need to learn relatively few things and become expert at them. You would essentially dictate what technology could and could not do for a business, and the business users would generally have to live with your choices.

As technology has become seemingly simpler, the tables have turned. "Credit Card IT" has begun to take root. Business users are evaluating technologies and purchasing them directly without asking your advice. You no longer instruct users about their tools and how they work. Not only are you no longer in charge, you are also overwhelmed with making a broad range of technologies work, while still having to ensure you maintain control over potential breaches in security and disruption to the business that may result from technology failing. The scale of complexity that you are faced with gives you less and less time to focus on learning a single technology set in depth.

Simultaneously, services usually provided by systems in private business-owned datacenters are being offered for easy consumption by a diverse range of vendors such as Microsoft, Amazon, Salesforce and Google. In effect those companies are commoditizing extremely complex solutions and in essence becoming utility providers. While you are attempting to learn about the new world of IT as quickly as you can, big IT vendors' utility approaches are undermining the need for you to do any of the work at all!

This leads to enormous uncertainty, and much of this uncertainty stems from traditional, strict silos of responsibility that are thrust upon different roles. For you to serve the business in the new world of IT, you need to be able to extend beyond the bounds of management and operations, and into the business realms. You need to understand the efficacy of a wide-ranging set of technologies and solutions and be able to make competent and credible decisions about how to deliver against the service needs of a business. This requires businesses themselves to determine the appropriate levels of stability and dynamism that should be present within their work environments. Perhaps you will need to evolve and become part of the business, rather than sit in a separate department. Perhaps business users need to get the new age of IT workers, like you, engaged more closely at the outset, rather than making credit-card IT choices without you.

Rita Gunther McGrath, a Professor at Columbia Business School, talks about waves of advantage in her book, *The End of Competitive Advantage* (Gunther McGrath R., 2013). She goes on to discuss the types of environments that become necessary for efficient operation of a business, including a combination of stability and dynamism.

TABLE 4.1   Stability and Dynamism in the New World of Work

Stability in . . .	Dynamism in . . .
Leadership	Talent Allocations
Strategy	Budgeting
Values	Business Portfolio
Talent	Individual Job Assignments
Customer Relationships	Decision-making

TABLE 4.2   A Transition to Individual Skills

From: Organizational systems	To: Individual skills
A stable career path	A Series of 'gigs'
Hierarchies and teams	Individual superstars
Infrequent job hunting	Permanent career campaigns
Careers managed by the organization	Careers managed by the individual

She continues on to discuss the impacts these shifts are having on employment criteria and employees. Some of those are summarized in Table 4.1. As the next generation of workers, you will probably need to view your employment opportunities very differently than your more seasoned counterparts close to retirement might. What used to handicap career growth has become an advantage. At the heart of it is the ability to cope with change. Your generation of IT worker will need to find stability in developing strong networks, and will need to be increasingly open to "gigs," unlike your predecessors who may, or may not, adjust to this mentality.

Sarah reflects on her own career to illustrate how remnants of the past are fueling the practices of today.

> Many years ago I used to work on very technical engagements. In my twenties I used to tell people that no technology worried me as long as they gave me 24 hours to learn it; and by that I meant a day, not three working days! Today the new generation looks to "gigs "similarly. It fascinates me how optimistic and alive they seem even as they engage on contract after contract, rather than seeking permanent employment. I wonder how they apply for mortgages and commit to long-term finance and other contracts without a reasonable view of where their income may be derived. The truth is that they actually have more stability in knowing that they are adaptable and can take on just about anything as long as they maintain their networks, both for expert assistance and for employment opportunities. They are not quite so shackled to an employer and desperately holding onto gainful employment out of fear that they might not find other work. Their energy is inspiring.

However, old norms continue to persist. Different industries change at different paces. For example, that which may hold true for the rapidly changing IT industry may not hold quite so true for employment in the banking sector or public institutions. Again Sarah expands on the topic with an example from her own career.

> When I was studying electrical engineering at college one of the lecturers informed us that technology doubles every five years. At the time it seemed to be unimaginable that we'd have to learn double the stack of textbooks we'd been given every five years. Of course it was a naïve response to his statements,

and no one really considered the skills one acquires on the job and through incremental learning, but the class all signed up for it. It was okay to look at the ever growing mountain of knowledge and want to learn more.

Today the problem is greatly exacerbated. The numbers of products, services, competitors and technologies you are exposed to on the global stage are enormous. You can't possibly acquire enough expertise to work on everything. It is with an air of foreboding that you may start to exhibit some odd behaviors. Some might stick their heads in the sand and pretend that the threat is not present.

Others of you will look to reframe things within the context of your past belief systems. Those belief systems may be as old as the ink on your diploma or a decade or two into your career. Inevitably you will start fighting back against the tidal wave of change, while doing your best to appear like you are all for change. It's an odd way to fight back, but you only need to look to history to recognize that those who feel the most pain during change are those that tend to cling onto what they have for dear life.

You could easily end up like many of the colleagues with careers fading around you. At first it's imperceptible, but in time as the lumens begins to reach critical lows the realization overtakes them. What could have started as a slow-paced and measured, yet still inexorable march into their future careers rapidly has become a wild sprint culminating in a leap of faith that many are unable to make. The aftermath of mass leaps of faith are rarely pretty. There are those that survive and prosper, but the vast majority simply is not able to bridge the divide. They cling onto what they have as long as they can, but eventually become obstacles to the path of progress. It is at that point that layoffs begin, and those who were unable to bridge the chasm find their careers in tatters as they take on lesser jobs, or similar jobs within slower moving sectors. In the technology industry, disruption will occur multiple times during your career and that trend is only bound to accelerate.

In their book, *Race Against the Machine* Brynjolfsson & McAfee (2012, p. 73) discuss this phenomenon in economic terms as they relate to employment and technology. The situation looks very bleak for those who refuse to adapt. Towards the end of the book they make apparent some of the potential opportunities open to you, including the delightful and thought-providing note that, "information doesn't get used up, even when it's consumed." They then go on to expand with, "the economics of digital information, in short, are the economics not of scarcity but of abundance" (Brynjolfsson & McAfee, 2012, p. 73).

Their writing represents a fundamental shift in thinking. It's a shift from protectionism and scarcity to opportunity in abundance. This is the type of thinking you need to instill in your career in technology. You have no choice but to start thinking differently about your career and employment with each disruptive event. Alternatively you can attempt to think and act proactively, planning ahead of each disruption in order to leap to the opportunity.

In essence, your chosen career path in technology includes a commitment to change and disruption. It's always been the case, but something seems different this time. This time it's not purely a technology problem. You are not only leaping away from one technology vendor and betting on another. The ideology is different too. Some of your peers will look to the changes needed and begin to feel like dinosaurs doomed to extinction while others might feel a sense of foreboding but proceed into the darkness regardless. It is for the latter that we're writing this chapter and we hope that is you.

In an old Microsoft game, *Age of Empires*, we sought to expand our empires into the unknown, tile by tile, and exert control over the environment. Your career will not be much different as you find yourself taking steps towards new opportunities tile by tile. Just as in the game, you will venture forth and find solid footing and the area around you will light up. You will regain your composure, just a little and survey the new landscape, and then plot your next steps forward. You know where you are going, but you don't quite know where it is just yet. You are a pioneer finding new worlds, but you are also used to living in a world where you are accomplished and viewed as an authority and you have your existing, but ageing, networks that you depend on.

Some elements of your potential future careers in technology are starting to take shape, but most of those changes are related to the technologies themselves. Recognize with increasing clarity that basic work, such as datacenter and foundational level work, is essentially disappearing as a plethora of IT vendors offer solutions that they control entirely. In the shorter term finding temporary security in jobs related to IT infrastructure will seem relatively easy. However, at some point as migrations to cloud infrastructures are completed a great many people will be fighting over the very few jobs that remain. Automation and machine learning are further transforming the employment landscape too. To some, who have depended on a smaller set of core skills for decades and earned a considerable income too, the future may look bleak. Use your knowledge of trends to define your next steps and ensure your relevance.

Coupled with an even greater seismic shift in the culture, Millennials (Generation Y, as defined by Pew Research Center as those being born after 1980) are projected to occupy approximately 40% of the workforce, and they think and operate with a very different philosophy and without the same perceived constraints of their predecessors in the workforce. Their views of long-term employment, also known as "a job for life," are markedly different. In short they do not expect to find "jobs for life" at all. They're not loyal to brands, and that includes the devices and technologies that they use. Technology has become so ubiquitous that they no longer look for which technology is simplest, but rather for those sets of technologies that allow them to be entertained, to communicate and to get work done as simply as possible. They expect those same technologies to be available to them in the workplace, and they bring that mindset into the workplace.

While the present IT industry shapes itself around Gartner's Megatrends, or at least some derivative of Social, Mobile, Big Data and Cloud, Millennials do not seem to be interested in them at all. To them it's all about experiences. They're building networks, developing skills, learning, and looking with fresh eyes on the opportunities ahead of them that all these amazing technologies provide. While they're doing so, the aging technology workforce is looking to incremental changes as their savior and hoping that they can iterate quickly enough. Seldom can people shackled to the metaphorical ball and chain that is their current skill set keep pace with those that are uninhibited! Sarah relates her own experience.

While I'm not an expert in psychology or education, I am a stakeholder in the development of a sizeable population of the workforce at my company. The problems are startlingly apparent as I experience both Millennial and Generation X employees. It's an interesting mix, especially as some from each generation are able to make the leap pretty successful, and others in the younger workforce have been, in a manner of speaking, indoctrinated through their education or circumstance to adopt older thinking. Coupled with that dynamic, we must also consider how we transfer knowledge and wisdom from one generation to another. The IT industry cannot flourish if it must first cope with the next generation of employees failing to learn important lessons from the past. Humanity doesn't change at the pace of IT. In a world where fast changing IT is also becoming a utility service, much like telephony and cable television are today, we simply can't bear the burden of technologists that do not take into account the average person. In the past the average person has depended on people with IT skills within their circle of friends and family, or paid support providers. They are far less tolerant of IT bungles now, and are much more adept at changing from brand to brand, interface to interface, device to device and operating system to operating system.

There are two theories that will help you as you embark through the various stages of your career. The first is the rather well-known Abraham Maslow's Hierarchy of Needs (1943), and the second is Sources of η Achievement (McClelland, 1961). While you may view Maslow's Hierarchy of Needs as hierarchical, try to think of it as a tool to assess what needs to change within your own metaphorical tool kit. It is also not necessarily useful to assume that its composition is hierarchical. For example, a highly accomplished and visible IT Architect that has achieved wide recognition in the industry may be looking to rebuild in some areas, but job insecurity may not end up affecting his Safety, as described in the hierarchy. Perhaps it's

Self-Actualization and some elements from the Theories of Achievement that need shoring up instead. David McClelland identified three needs that seem well-suited to the changing landscape, namely the needs for achievement, power and affiliation. When extrapolated they are easily understood. Those who are motivated to achieve will seek out difficult and complicated challenges. Those motivated by affiliation tend to collaborate and to mitigate risky, complex and difficult work by building teams and partnerships to undertake them. Those motivated by power may be motivated by status, or feel a deeper desire to encourage and teach. Of course people may exhibit various combinations of these traits too. Sarah reflects on her own experiences.

> For example, I enjoy connecting, enabling and empowering others. At face value people might want to assume that means I'm motivated by power, but my way of working is collaborative, and I prefer to be the person in the wings that challenges and pushes great talent to be all that they can be. I take on the most difficult challenges I possibly can too. In essence I am motivated by all three to some extent or another. When I look to the shifting cultures I can see the balance of the motivators shifting too.

The scale and complexity of the challenge cannot be underestimated. There is no proverbial silver bullet or one size fits all approach. At present the best approach seems to be technology centered combining some ingredients of the new characteristics of culture with new job roles that describe the technology requirements differently. The IT industry as a whole is not seeking specific operating systems and application expertise for positions that are helping customers with strategic approaches. These skills have become basic "table stakes," rather than the differentiators you may have viewed them as in the past. Instead, the search is for people who can do work related to what is termed in the IT Architecture world as addressing business requirements, cross-cutting concerns and quality attributes; essentially those elements that cut across all of the technologies and solutions being implemented, rather than a specific technology only, i.e., "social" experts. There is no one definition of "social" that suits all vendors' implementation of social technologies, nor is there one simple definition of "social networking" that adequately describes the expectations of those that want to harness those sorts of technologies and services. Social expertise essentially describes a broader cross-cutting set of skills that may include communication, collaboration, teamwork, change management, migration, implementation, management, operations, security, privacy, application, service and other expertise that may be related to well-known Internet services or not!

Other technology jobs, such as Customer Success Managers, Cloud Architects, Chief Digital Officers, Data Scientists, Cyber-security Consultants, Value Realization Executives, and Experiences Experts who may focus on high value experiences are fundamentally different in both title and description from those traditional roles that you may have encountered before. It is easy to become trapped into thinking about these roles from older viewpoints and simply state that there is nothing new in them, but to do so would be a mistake. One cannot simply apply a technology lens to these new jobs.

These jobs of the future incorporate sweeping cultural changes that are overtaking the IT industry far more rapidly than anyone ever anticipated. When you hear about attributes such as communicative, collaborative, decisive and nimble, an expectation is being set for a workforce that is essentially the counter-culture of that which prevailed before. Talented jerks, empire builders and those who hold onto knowledge as a means of power are quickly becoming the dinosaurs of the IT industry today. As the culture shifts, all of their clandestine maneuvering and attempts to disguise their bad behaviors by saying the right things have become so transparent it is clear their tenures are coming to an end. The move to change started decades ago captured by a few in the hobbyist approaches of those that founded the post-mainframe era and those in academia who simply shared source code long before there was an open-source movement. Those hobbyists and open-source enthusiasts were in the minority. Over time, the small little movements gathered steam and momentum, and today it has resulted in massive cultural shifts that no company can choose to ignore. Some of their attributes present themselves today, but their insatiable curiosity, partnering to explore and innovate, and their open sharing of new discovered and acquired knowledge have become a kind of new normal as we witness the pitched, almost religious fights, about patent law and copyrights.

As we watched the 2013 movie *The Internship* (Levy, S. Director. 2013. United States: Twentieth Century Fox Film Corporation) it became immediately apparent just how different the appeal of teamwork in a business setting was to the younger generation. We had the opportunity to watch it with a group of younger people who all wanted to work in a setting like that. Oddly it reminded us both of an earlier time in the IT industry; a time when we first joined it. The curiosity, the passion, the long nights of "geeking out," the thrill of engaging in a shared endeavor, and learning and succeeding with others were all ingredients we lost in the IT industry as it became more formalized. We're receiving a stark reminder for Millennials that these ingredients remain critical to our tenure and success in such a forward thinking industry today!

Somewhere over the past twenty years we lost what made the IT industry special, and it has taken younger, successful companies to remind us that we're not a staid, stuffy set of institutions trading in goods and services that people have needed since the dawn of civilization. A large part of the industry is aspirational and is focusing on what comes next. Amidst optimization, efficiencies and "bean counting" (accounting) the essence of IT has faded, and it's taking the very real threat of future irrelevance for people to recognize that jobs and expertise need to constantly grow and grow and grow.

This is where you will (re)start anew. Some will become overwhelmed by the darkness as their careers fade out of existence. Others will take the leap required into a bright future. They will be filled with anxiety, uncertain of what lies ahead and hopeful that it is not for naught. In our chapter we represent this new future through the metaphor of a white room.

As you enter the bright light of a seemingly barren, white room, you can look to it as Room 101 from George Orwell's novel, *1984*, or you can approach it with all of the knowledge, wisdom, foresight and curiosity gained by the momentum of those that preceded you. You may enter the future with little more than what you can carry, but it will be an irreversible step out of the darkness and into the light. Eventually you will compete on a level playing field with many people who are much younger than you, knowing that embracing the new world that they introduce is that next big wave to catch in order to succeed.

Let's take the leap into that room, and your future.

A NEW BEGINNING

It is as if you have woken up in an empty room. You are the only person within this cube where the four walls, the ceiling and floor are white. You assume it's a room, but maybe it is a box. What is your first reaction? Do you think about what is not there? Or, do you think about potential, what could be there, about possibilities?

It seems like just yesterday you were in class with your friends, playing corporate games and scenarios and you had all the answers. You knew what to do and how to change the world and could not wait to get started. But now, you are in this room with nothing but a laptop, a phone and whatever you have on your person.

This is not simply a glass half full or empty question, and it is not limited to positive or negative attribute based thinking. Innovation requires all types of thinking styles in order to reflect the variety of consumers related to the end solution. It is all about how you approach this box, this cube, these four walls and the emptiness—or fullness of emptiness.

Is this a social experiment to see how long you can survive in isolation, in solitude? Why is it so quiet? Why is the light so bright? Are you being watched? How did you end up here—this is not what you expected.

If you are limited by the four walls of this scenario and cannot see opportunity, you are likely limiting yourself in other places and ways in your life also. Not to be cliché, we have all heard thinking of outside the box.

What about thinking inside the box? In this case, inside this room? How do you take goals, constraints or boundaries and translate any of these into creative resolve? The reality is that it is extremely rare for any project to have no boundaries. There are always constraints or preconceptions polluting projects prior to inception.

Back to the box. No one has told you what to do here, so the sky (or in this case the ceiling or the lid) is technically the limit. Consider the psychological perspective, in this case the *Hierarchy of Needs* (Maslow, 1943, p. 370). Maslow was a behaviorist that believed people's actions were driven by needs in a hierarchical and weighted manner—a pyramid (see Figure 4.1). The lower in the pyramid a person's needs were, the more desperate the action to resolve the need would be, in some cases resulting in acts of violence or desperation (primal need).

This box is an unusual scenario in terms of Maslow; you have physical shelter; some of your Basic Needs are met. The temperature is perfectly controlled. There is no visible source of food or water, but let's assume that if for some reason you were to exhibit hunger or thirst in the room your need would be fulfilled. For the moment your Basic Needs have been met.

If you were in or felt you were in a precarious situation in the room, your primal instincts would be outweighing logic to overcome fear. The decisions that you make would not necessarily be logical or sustaining. Time would be a driving factor and survival would be the primary goal.

As you sit in the room, you begin to hear the loud ramblings of humans coming from the outside the walls. Is this a threat?

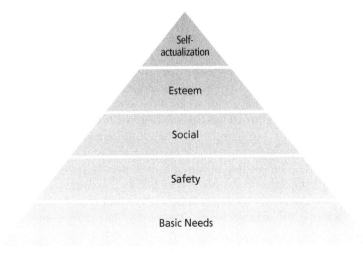

Figure 4.1   Maslow's hierarchy of needs.

There is a door across the room from where you are sitting. You stare at the doorknob, contemplating walking up and turning the knob to assess the crowd. But you are stunned and sit statuesque in your chair. The noise continues.

You begin to fumble through your pockets. How can you protect yourself from the potential risk? No one has come through the door yet, no one has fumbled with the door knob—maybe you are safe?

Assuming all is well, you quickly pull together a few things from your pockets to see if there are potential weapons to be mashed-up or fabricated. Your credit cards can be broken to produce sharp edges to act as poking objects, the inhaler in your pocket can be used to blind someone with a simple aim and expel and your belt a whip and so on. Fear is a motivator and drives creativity but not necessarily efficacy. In fact as the noise subsides and your heart rate calms, you realize it is not likely that most of these tools you created would have succeeded at fending off the previously perceived approaching hoard. You laugh off the "weapons" you have constructed and relax back into a slump in the chair again. Revisiting the old adage: Necessity is the mother of invention—we have learned that as a mother, it appears to yield fewer successful inventions than might be required to survive.

This first attempt at productivity is reactive and can be likened to the Storming stage of the Five Stages of Group Development (Tuckman, 1965). You assumed that you sought security, and in a state of anxiety decided you needed weapons to protect yourself. Your assumptions may end up preventing you from finding others to collaborate with, and perhaps that may be the most likely path to extricate yourself from the box. Storming is described as a time of feeling uncertainty, fear, and exposure to potential harm. These are all feelings that cause resistance to the soundness of ideas and decisions. While there may be success in this stage at identifying an innovative solution, there will be more failure than success due to the influence of emotions.

Now that things have settled, you assume you are safe in the environment. The walls were not penetrated by the mob or any other external elements. It is time to aspire to the next rung of Maslow's pyramid. Without immediate desperation, you have the privilege of thinking things through vs. the 'fight or flight' biological response experienced earlier.Silence follows. Eventually you begin to notice every sound of your existence: the tick of your wristwatch, the rustling fabric with each breath or subtle movement and the sound your eyes make when they blink. This almost anti-sound of being that you ignore any other day is now the center of your universe.Time to take inventory. You take off your sweater and kick off your shoes. Emptying pockets, you check your phone only to find limited bars. You unwrap your laptop. Before using the electronics, you need two things—electricity and connectivity. Searching the walls you find that there are masked outlets, as you plug in your laptop to test the current, you see the charging symbol. You power up the machine, and follow the instructions on the piece of paper tucked inside the flap.After a few attempts, you are able to connect to a wireless network. This is the turning point. In modern terms there is no feeling of sanity, safety or basic survival without a phone or laptop and connectivity—you have this. You are good, you can do anything.

The machine is loaded with various programs. You click on the email icon. There is one ominous email in your inbox. Today's tasks. You follow the list clicking on links, registering on sites, downloading software—exhausting. Hours later you are done. You lean back in your chair feeling accomplished, but when you review the list and assess the emptiness of your cube you realize that you really haven't done anything at all.

Now you are able to seek out others. You attempt to send a text to someone in your directory—it fails. You then attempt to connect to a few social networks, none of the web pages are accessible. A deep sigh, loneliness is not an option.

Another email arrives. Subject line: "Next Steps."

A long numbered list of tasks with links, expectations and timelines. You pick up where others have left off. You add your part; you pass it on to the next. You fall into a pattern—according to Tuckman, you have Normed into a steady state.

Today turns into days, into weeks, into months and a year passes. You realize that all of those aspirations you had weren't real. You are just a cog in an invisible machine. You are a number and you are not making an impact, you are not changing the world. You are just sitting at this desk and writing code. This is Performing (Tuckman), this is not what you signed up for. If you don't do something, it will go on and on with no end in sight- adding your code to the pile like a factory worker in an assembly line with no real connection to 'why.'

CONCLUSION

In this chapter we attempted to blur the line between science and art as we analyzed today's workplace in terms of two methods of approach. The first was the seasoned professional aspiring to maintain dignity and relevance in an exponentially changing world of technology. The second was both a new college hire thrown into the bullpen of coders in a cube farm of entry level technology jobs, and a seasoned professional who finds himself in a similar situation as he takes on new tasks in a radically changing career. While each has individual woes, they are similar in their efforts to ascend Maslow's Hierarchy of Needs pyramid in order to achieve their own interpretation of basic needs, security, connectedness and contribution towards self-actualization. Each is pitted against the other, threatened by the other. They are both tied to the realities of pioneering tomorrow as they challenge job scarcity with new creative approaches to design resulting in solutions to fit their needs or desires or simply to fill their pockets when other options are not available.

It is no longer about glasses that are half full or empty. Relevance is achieved through an ever-changing set of hard technological skills, where the possessor of the biggest toolbox of soft skills will come out on top. There is no 10, 20, or 30 year career plan. It is likely that your tenure in a company will be in the single digits for a large part of your career, unless the next generation forces a shift in this paradigm. Challenging every action, step, move are these overwhelming questions:

How do you prepare for emerging careers that are barely defined?

How do you put yourself in a position to understand technologies that are barely a figment of someone's imagination, let alone born?

How can you identify skills to develop that will last past your first paycheck?

In the world of IT Architecture, people depend on identifying and harnessing patterns in order to solve problems. These are derived from what is observed by problem sets being solved more than once, and usually incorporate the learning extracted from failed attempts. With the wisdom of hindsight, we work to distill the common patterns used to achieve resolution and we proceed to use, re-use and, perhaps, overuse these patterns. Eventually the consistent application of patterns continues to solve the problems that they're designed to address, but we no longer require the same wise pioneers to implement them.Repetition of patterns that were once deemed creative, exciting and unexpected becomes mundane. They become perceived as cliché or corny, and are treated like remnants of a bygone era even though they deliver enormous value. In other words, take history and the related lessons learned with a grain of salt. We urge the application of reason rather than taking patterns as prescriptions to the future. In order to prepare for the future we can look to the past. The history of technology will not necessarily repeat itself, but can provide breadcrumbs that lead to the future.

In order to remain relevant, you will need to remain one step ahead. However, that alone, is no longer enough! Having the presence of mind and the ability to contribute to the next commodity offering a short time before others might is no longer ascribed the same level of value as it once was. You have to develop a certain "street cred" before you become the thinker driving the mass of coders in cubes versus being one of that mass following the orders of the thinker. "Street cred," or reputation, will help to establish your standing within a network of people.

We leave you where we started, a desire to fulfill your basic needs. According to Maslow—your basic needs and safety will always trump self-actualization. We feel you need to break out of this paradigm and embrace what we might have perceived as risk taking if you're to avoid ending up in a four walled cube slinging code amongst the masses waiting for the next task from a "thinker." When the need to quell that unspeakable din of the empty room eats away at your soul you have two choices—carry on, or break free into the wilderness and venture into the new world where incredible feats of innovation and creativity lead to a new form of success.

REFERENCES

Brynjolfsson, E., & McAfee, A. (2011). *Race against the machine: How the digital revolution is accelerating innovation.* Lexington, MA: Digital Frontier Press.

Gunther McGrath, R. (2013, June 4). *The end of competitive advantage: How to keep your strategy moving as fast as your business,* Harvard Business Review Press.

McClelland, D. C. (1961) *The achieving society.* Mansfield Center, CT: Martino Publishing.

Maslow, A. H. (1943). A theory of human motivation. *Psychology Review, 50*(4), 370–396.

Tuckman, Bruce W. (1965) Developmental sequence in small groups. *Psychological Bulletin, 63*(6), 384–399. American Psychological Association.

PART III

EVENT HORIZONS

CHAPTER 5

NAVIGATING THE WORLD OF INNOVATION

A Suggested Path for Today's Business Managers

Min Basadur
McMaster University

ABSTRACT

How might leaders stand out from the rest, especially in a turbulent economic environment? Not by spouting the same efficiency optimization skills taught in most schools, but by mastering adaptability, a seamless creative process combining analytical and imaginative thinking skills. This tangible, proven process encourages deliberate and proactive innovation. Concrete real-life business cases are provided throughout showing how business managers can distinguish their units by recognizing, measuring, and collaborating the design thinking and problem solving styles and preferences of their staff and team members through a four-stage process of problem finding, problem definition, solution optimization, and implementation for innovative edge.

Millennial Spring: Designing the Future of Organizations, pages 87–105
Copyright © 2014 by Information Age Publishing
All rights of reproduction in any form reserved.

Many things will be different by the middle of this century. Managers will be collaborating, thinking and problem solving at more innovative levels. As technological advances in social networking, big data and artificial intelligence provide more insightful information and more reliable analytical tools, tomorrow's managers will differentiate themselves through their design problem solving. With technological advances available to all competitors and offering only a temporary competitive edge at best, it will be the cognitive and collaborative skills of managers that provide real competitive advantage. Effective use of large amounts of information and sophisticated decision support systems will require managers to be skilled in articulating the right questions to ask, and the correct queries to make. Rather than commandeering hurried hit or miss solutions, they will learn to save time and increase accuracy by following Einstein's wise saying: "If I had an hour to save the world, I would spend 55 minutes defining the problem and I would need only five minutes to solve it." Rather than delegating solutions they have created themselves, tomorrow's managers will need to be adept at handing off fuzzy problems to well-designed teams skilled in fact finding and problem definition, as well as the creation of solutions capable of attracting the necessary consensus for implementation. In a world that demands innovation, the ability to integrate analytical and imaginative thinking—and inspire it in others—will become increasingly essential to success. With that dream for the future in mind, I share the following thoughts and real-life lessons on innovation, creativity and team building.

A decade ago, I received a phone call from a former client named Barry. He called to ask if I remembered leading a creative problem solving session he had participated in with a group of about 10 others. Now living in New York City and working on an advanced degree at the Pratt Institute of Design, he told me the session had helped a group of designers move forward on a city project and said it had made a major impact on him. "Designers think they are pretty good problem solvers, but the process you used taught us how to problem solve way beyond what we thought we knew," he said. "Designers seem to have their own special way of solving problems but can't put it into words. The process we used helped make design thinking more explicit and took it to a higher level."

Now, the term "design thinking" was new to me, but I was glad to come to understand it as a creative problem solving process like the process I have taught for many years rather than some kind of mystery for a select few. The process pulls together a person's analytical and imaginative thinking skills, and tries to simplify how to apply creativity at work. Along the way, I have come across a wide range tools, techniques, and philosophies offered analytically trained people to perform more creatively. They have different names and are described in a variety of ways. This scattering appears to have made it confusing for people genuinely seeking the skills that will help them distinguish themselves in today's rapidly changing business world.

In this chapter, I hope to simplify and explain the concepts that underlie the process of creativity, to help managers who may be struggling to transition from the "old way" (sometimes called the "manufacturing economy") into the "new way" (sometimes called the "knowledge economy"). While the need for adaptability is rampant, skill in developing it is in scarce supply. By driving innovation in day to day work, twenty-first century managers have an opportunity to accelerate their careers, "stand out among the rest" and differentiate themselves from their peers, most of whom are equipped with the same basic credentials. To do this, they need to master a structured process which simplifies and enables creative thinking to drive innovative results. Developing skill in a process that combines imagination and ambiguity with analysis and certainty will provide them with the base for building adaptability within their organizations.

Balancing Efficiency and Adaptability

We live and work in an era of rapidly accelerating change with frequent discontinuities and interruptions. Many organizations that prospered during more stable times—times that rewarded routinized efficiency—now find themselves poorly adapted to today's new economic and social realities. Everywhere we look, traditional structures are abruptly being reshaped or falling down. Once successful companies are finding that their sure-hit formulas no longer work. Long revered icons of organizational excellence have been humbled, and even bailed out of bankruptcy and imminent demise by government intervention. Organizations whose main virtues during previous times were predictability and reliability are finding it difficult to adapt to this increasingly dynamic environment.

In previous decades, the role of the manager was to improve efficiency and maximize the next quarter's results. Those goals required routine based, analytical thinking and decision making—skills widely taught in most business and engineering schools and universities, and rewarded and reinforced in most organizations. But while still valuable, an organization's efficiency is now recognized as only one half of the formula for success in today's shifting economy. The other half of the success formula demands that organizations develop adaptability skills (Mott, 1972; Dolata, 2013; Basadur, Gelade, & Basadur, 2013).

Efficiency means perfecting routines in order to attain the highest quantity and quality for the lowest possible cost. On the other hand, adaptability means continually and intentionally changing routines and finding new things to do and innovative ways to do current work. Adaptability means scanning the environment (Simon, 1977) to anticipate new problems, trends, customer needs, opportunities, then deliberately changing methods in order to attain new levels of quantity, quality, and cost and new innovative methods, products and services. To develop adaptability and build competitive edge, managers must expand their thinking to include non-routine based, imaginative creative thinking and problem solving skills.

After providing Barry with some training in Simplexity Thinking, I visited him in New York City and spoke to senior partners at the advertising and communications firm where he worked. They told me that their clients were increasingly uncertain about how to continue to grow their businesses. "In the old days, General Motors would come to us and say, 'Here is what we want, go do it. Run some focus groups and send us the results'. But nowadays, they often ask us what they should be doing. Our people are having trouble coping. They used to be able solve problems well. But nowadays, no one gives them a problem. More often our customers are looking for help in discovering the right problems to solve."

The firm had discovered that solution formulation was no longer the name of the game. Instead problem formulation—made up of problem finding and problem definition—was the key skill set in their workplace. They also discovered that problem formulation skills can be improved through training, and through the use of a clear, delineated process and helpful tools.

The importance of problem finding is becoming more widely recognized (Kabanoff & Rossiter, 1994; Short, Ketchen Jr, Shook, & Ireland, 2010). Circumstances are changing, including all kinds of technology advances, while people are changing in terms of demographic trends, education levels and connectedness to others. Innovative new products and services are required—and quickly—to stay ahead. Top companies are discovering that innovation begins with problem finding—discovering customers' (and potential customers') problems ahead of the customers themselves, then providing solutions in the form of new products and services (Basadur, 1992). Problem finding, which is the first step in innovation, begins with listening to your customers asking the right questions. The most important parts are not the answers, but the right imaginative questions. Beginning with the imaginative questions: "How might we? How might our customer?" Exciting challenges can be found when companies ask themselves questions like, "Why might we? Why might our customer want to? What might be stopping us? What might be stopping our customer from?"

Toshiba Corporation told me, "When we hire new scientists and engineers, we keep them out of R&D for two years. Instead, we place them into the sales department to begin their careers. We want them to learn that their job is to learn the problems of the customer. We want them to know we are not going to hand them problems to solve. We want them to know that innovation begins with finding problems to solve." In innovative companies, the word problem is synonymous with need, challenge, want, desire, opportunity, puzzle, change, trend, observation and many other triggering terms.

The Process of Innovation

Innovation is a process, not an event or an outcome (Basadur & Gelade, 2006). It is a process of finding and defining internal and external customer needs, developing solutions to address those needs, and successfully implementing those solutions. The needs—or problems to be solved—can be found across a broad spectrum of areas, including, but not limited to technology, products, markets, packaging, design, manufacturing processes, new business models, and new ways to go-to-market. The innovation process and the mental skills that make it work can be learned and become a daily habit that results in ongoing creative disruption and problem solving (Basadur, Graen, & Green, 1982). Everyone can take part in this innovation process. Once learned and understood, it can be used in every department and by people at every level of an organization (Basadur, 2013).

In an Executive MBA course I teach, I help my student managers discover that they must internalize the process of creative problem solving. It must become part of everything they do every day; part of their vision of the world. Afterwards, the students emerge very changed, and say, "This should not be an elective course; this should be the first course everyone takes in business schools. It teaches us *how* to think, not *what* to think. It is the basic learning in the process of management. All the other courses are a bunch of content subjects that teach us what to know (finance, accounting etc.) and can be taken later and fit into the process. These subjects are important to know, but need to fit into a process of creative thinking and problem solving that guides all of our activity."

It is essential for all of us to recognize the importance of a creativity process to help us navigate our way to prosperity in a very turbulent world. While managing and improving efficiency is still an important skill, it is only one side of building a successful organization in a shifting economy. Today's managers must also expand their thinking to include the imaginative creative problem solving that builds organizational adaptability (Basadur, Basadur, & Licina, 2013).

We've traditionally explained creative problem solving as a continuous and circular four stage process that begins with the deliberate seeking out (generating) of new problems and opportunities (see Figure 5.1). The second stage of the process is conceptualizing, which involves formulating, defining, framing and constructing a newly generated problem. In the third stage, problem solving, evaluation and selection of solution ideas

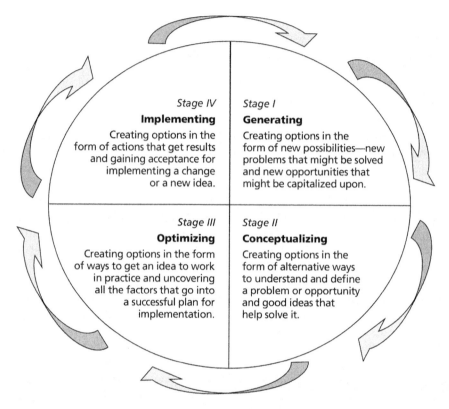

Figure 5.1 The four stages of the simplexity thinking process.

takes place, while the fourth stage results in solution implementation. The process then begins anew, as every implemented solution (action) results in the opportunity to discover (generate) new problems and opportunities (Basadur, Basadur, & Licina, 2013; Basadur & Gelade, 2003).

While the first two stages of the Simplexity process tend to be more imaginative, and the last two stages more analytical, the ability to think up options (diverge) and the ability to evaluate options (converge) is used in *all* stages. Skilful use of the process draws upon a variety of kinds of knowledge, disciplines and expertise.

However, Barry's observation about the relationship between Simplexity Thinking and Design Thinking led me to consider other ways the process could be described. Some time ago, psychologist William J. J. Gordon suggested that inventing and learning are opposite forces which feed each other in turn (Gordon , 1956; 1971). Inventing is characterized as a process of breaking old connections. Learning is characterized as a process of making new connections stick. When we invent, we "make the familiar strange" (by breaking old connections which compromise current understanding). This permits us to view old phenomena in new ways, although this can be uncomfortable at first. When we learn, we "make the strange familiar" (by making new connections between new (and thus strange) phenomena and our current understanding. This permits us to view new phenomena more comfortably.

In the circular depiction in Figure 5.2, the problem solving process is viewed through this perspective. On the left side, new "paradigms" (ways of thinking and doing) become established. New processes are learned and become well-known and comfortable habits. On the right side, such old established paradigms are broken. New processes that produce better quality or new goods or services are invented to replace previous processes. When an old familiar paradigm such as a well-established business process is broken, the new one replacing it feels very strange and uncomfortable to everyone affected. They are experiencing a process of unlearning, breaking connections with past understanding and letting go of old habits and beliefs. As time goes on, the new process becomes less strange, and more familiar. This is a learning process—making new connections and adopting new habits and beliefs.

This cyclical process can also be viewed as representing the 'operating' versus 'inventing' sides of the modern business world. As new 'inventing,' or pattern breaking activity occurs, old and familiar processes are transformed into new and unknown activity. As we travel around the ongoing

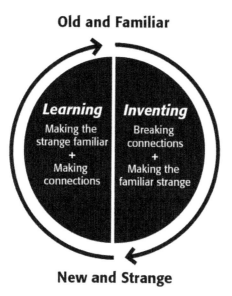

Figure 5.2 Two halves of simplexity thinking: A continuous process of inventing and learning.

circle, those new patterns are converted into new familiar processes, and readied for additional transformation.

However we name or describe this ongoing process, it is the basis of adaptability and innovation, and must be adopted as an everyday part of organizational life.

For individuals, internalizing the two sided process of innovation will result in a new circular pattern of thinking and behavior that will be evident as different, effective and innovative. The process will develop skills in seeking new opportunities for change (no matter how disruptive they may seem at first) defining and clearly understanding those opportunities, allowing new ideas to emerge and flow through the necessary steps of evaluation, analysis, testing and optimization until new solutions (products, services, or procedures) are created and step-by-step plans for implementation are developed and undertaken.

This process is not about coping with change. It is about *making change*. There is huge difference. *Change making* is real leadership in our new world. It is the game changer. Change is not something to be feared and imposed on people; change should be the result of proactive involvement of people in making new and valuable things happen to keep in the forefront, ahead of the rest. The beautiful thing is that our research has clearly demonstrated that when people are involved in creating change, they become motivated in all of their work (Basadur, 1992; Graen, 2014).

Organizations that recognize the value of breaking old and out-dated paradigms and replacing them with new and better ones, and actually know how to do so are what I describe as 'thinking organizations'. A thinking organization can both unlearn and invent. It is proficient in efficiency thinking (perfecting current routines), adaptability thinking (breaking old routines and creating brand new ones) and flexibility thinking (operating effectively when there are no routines to follow in ambiguous, unexpected circumstances). Thinking organizations engage the innovative abilities and creative aptitudes of all of their employees. (See Hazy & Backström's chapter.)

Unfortunately, few organizations have the skills or expertise to do this, often because they lack a framework for sustained and disciplined creative thinking. By adopting a structured innovation process, organizations can learn to think creatively in a collective, synchronized way, not only to improve routine work (efficiency) but also for the non-routine work of adaptability (See Basadur & Basadur, 2011, and Graen's chapter).

Solving Real Life Problems

Creative problem solving begins with problem finding, then moves into problem definition. This where "out-of-the-box" thinking is most likely to emerge, resulting in an unexpected twist or angle that leads to even more unique solutions for evaluation and optimization. When implementation of a new solution or product occurs, the process begins anew in a circular fashion, as the change that was made will inevitably alter things, (disturb the status quo) and result in new problems to be discovered. For example, the automobile's invention provided not only a new solution to an old problem (improving transportation) but created many new problems (e.g., infrastructure, pollution, energy and accidents).

Our traditional organizational approach to problem solving has tended to discourage an open-minded attitude. We are taught solutions or formulas we are to deliver when certain problems are encountered. Discovering how to abandon these habits is a challenge we all face in attempting to become more creative. For me, a turning point came in 1971, when I was asked to help a group of R&D chemists deal with a problem they were grappling with. I began by helping them leave what they knew and move onto creatively defining what they were trying to do, exploring the right questions to pursue, creating option al challenges beginning with the phrase "How might we?" and most of all, stay out of solutions. The challenges were then related to each other through the creation of a visual map that asked, "Why might we want to ... ?" and "What might be stopping us from ... ?" The group found this activity extremely helpful, with several new and unexpected insights emerging as challenges were reframed. I was amazed to see how new this type of thinking was for them. They were clearly unfamiliar with the key role problem formulation plays in the creative process (Basadur, 1995).

As word of my work spread, I was involved in many similarly and equally enlightening situations. I was asked to sit in a with a struggling product development team to observe and provide feedback on how they worked together. During their conversation, I heard the phrase, "I wonder what Andy really wants?" Who was Andy? It was the team's boss, sitting just up the hall with his door wide open. Why wouldn't someone just go and ask him? Because it had been three months since Andy had handed the project to the team: "Come up with another liquid cleaner." The company was already successfully selling two liquid cleaners, and the team seemed to realize it wasn't really sure what they were trying to do or why. Of course, no one wanted to appear stupid by telling Andy they had spent three months without making any progress. I then led the team through the creative problem solving process, beginning with problem finding, which we also sometimes call the 'fuzzy situation'. In this first step, we are either looking for a new problem or we have found ourselves in a situation which is ill-structured, ambiguous and undefined. Such a problem is sometimes found simply by being there or having a problem handed to us. If we are creative, we know enough not to assume that we already know what the problem is. Instead, the process calls for us to defer judgment and put our effort into moving from a fuzzy situation to a clearly understood statement (or family of statements) beginning with the challenge, "How might we …?" In this case, a key unknown fact that emerged was that Andy came to them three months ago after meeting with his own boss over lunch and gave them the third liquid cleaner project. What the group did not know was that at lunch Andy's boss had said, "We need to boost profits," and mused, "Perhaps we should add another liquid cleaner to the two we already have." Later, the boss did not even remember the casual conversation which Andy had taken as serious direction. When I led the team through the Why might we want to? and What's stopping us from? analysis, they reformulated the challenge to, "How might we create a new product that somehow keeps households cleaner?" When shared with Andy, the new challenge was heartily accepted as a major move forward.

Another insightful discovery about how people think or do not think creatively also occurred early in my career at Procter & Gamble. I was asked for help by a product development team that was formed at short notice in response to a competitor's new product. Colgate's green-striped Irish Spring was the first striped soap bar introduced to North America. With its aggressive advertising campaign emphasizing 'refreshment,' the soap was finding ready consumer acceptance. One of the rules at Procter & Gamble was that if we were the second entrant into a new market, a new product's competitive advantage had to be demonstrated prior to market testing. When I asked the team what was going wrong, they said that they had been unable to produce a green-striped bar that was preferred over Irish Spring in a consumer preference blind test. The team had experimented with several green-striped bars, all of which merely equaled Irish Spring in blind testing. It became evident to me that the team had chosen to define its problem as, 'How might we make a green-striped bar that consumers will prefer over Irish Spring?'

Applying the creative problem solving process to the problem began with developing alternative ways to frame the challenge. By repeatedly asking why might we want to make a green-striped bar that consumers would prefer over Irish Spring, we generated many alternative How might we? challenges. The flash of inspiration came from the answer: 'We want to make people feel more refreshed.' This led to the new challenge: 'How might we better connote refreshment in a soap bar? This less restrictive challenge, which included no mention of green stripes, gave us more room for creative solutions.

About 200 solution ideas for refreshment ideas were quickly diverged. On evaluation, two ideas stood out. One was an image of sitting on a white sandy beach with blue sky, white clouds, and enjoying soothing, cooling breezes. The other was based on travel to the sea coast for refreshment. The eventual product result was a blue- and white- swirly bar with a unique

Figure 5.3

odor and shape, which quickly won a blind test over Irish Spring, then soon achieved market success under the brand name Coast (Figure 5.3).

Solving this problem once it had been properly defined took the team mere hours. By leaping prematurely into solutions, the team had wasted almost six months before coming up with that problem definition. Many people and teams practice what we call "1 to 8" behavior. They skip the process and instead, jump directly from problem to possible solution, over and over again.

Successful problem solving requires people to begin the process with the recognition that they have a fuzzy situation and need to gather facts prior to defining the problem. Only after that is undertaken in a thorough fashion should they move on to exploring, evaluating and selecting solution ideas; planning for implementation; and finally, taking action. The step-by-step process is detailed in Figure 5.4.

As these examples illustrate, the first half of the creative problem solving process is the *more* imaginative half. This is where questions and inquiry are raised, problems are surfaced, hidden facts are discovered, breakthrough challenges are defined, and innovative solution ideas are hatched. The back half of the process is where the solution ideas are analyzed against constraints, evaluated and developed into practical solutions to be tested, perhaps including prototypes or drawings with step-by-step plans to gain acceptance by those impacted by the change. The final step involves putting

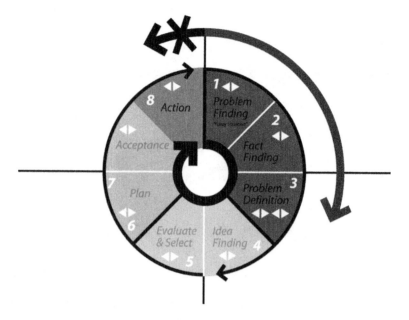

Figure 5.4

a solution into play, which requires overcoming the natural fear that comes with unfamiliarity and lack of certainty. (Naysayers will ask "How do you know it will work?") The experience gained becomes a new learning experience that launches a new round of innovation. The status quo is disrupted and new opportunities arise with new problems to solve.

Fear of being wrong causes other weaknesses in managing. It prevents important facts and challenges from coming out, which can result in the creation of solutions that are not on target and lack commitment to implement. I was once asked to facilitate a team that had been struggling for over a year to improve the efficiency of its potato chip shipments. The problem was that, on average, the trucks were travelling half full. Some key facts emerged during the fact finding step of our process. One was that an outside vendor had proposed a new way of loading the trucks which would fill them completely and result in an annual savings of $12 million. Another fact was that the team had been conducting tests across the country to check if chip breakage would be negatively affected by the new method. When I inquired about the findings, the team said results were inconsistent and they were now conducting additional tests. As well, the Market Research department had stepped in to study if there might be an optimum level of chip breakage that users might prefer, and were also getting inconclusive results. Which led them to run more tests. Everyone seemed to be going around in circles, in danger of running tests endlessly. To help the team focus and move forward, I engaged the members in using their creativity in problem definition using the what might be stopping our analysis. On a flip chart pad, I wrote the challenge "How might we write a recommendation to management by 3 PM today to approve the new loading technique?" The first answer to the "What might be stopping us?" question was that the breakage testing was not complete. The second answer was similar: they were not yet sure about the optimal chip breakage level for consumers. I noted these two challenges, then asked the question a third time. (This is how the process works. In every step, we push divergently to expand our thinking.) We waited and waited. It was painful. Someone finally said: "I think we are afraid to make a recommendation without being completely sure. We do not want to be wrong in front of management." This new fact, which had never before been clearly stated, led to a new third challenge: "How might we write the recommendation for change in a way that will explain the risk and ask management to take the risk with us?" The team selected this challenge and wrote the recommendation well ahead of 3 PM. It was immediately approved the next morning, with $12 million accruing to the bottom line. The team never thought that such an emotional fact would be legitimate to bring up in their work. In real world creative problem solving, fears and other emotions are critical facts to bring forward. Without acknowledgement of the human element present in all circumstances, proposed solutions are simply hypothetical exercises unlikely to result in success.

These examples serve to emphasize how crucial the problem definition step of the creative problem solving process is. Success in this stage often relies on our ability to overcome the habit of prematurely jumping to answers and solutions, and sometimes requires us to dig up hidden facts or admit to challenges people are unaware of or fearful of saying (Basadur, 1994). The transformative shift into a questioning mindset, which is accomplished by asking "how might we?" and "how might the customer?" is a reverse shift from what most people have been taught in schooling and culturally. This art of asking the right questions underlies the empathy and simplicity skills required in every stage of this process.

Leaders Build Adaptability and Creativity

Adaptability is a continuous change making process. In recent field research, top CEOs were asked for their thoughts on what good leadership was in the twenty-first century. Resoundingly, they first defined leadership as driving change, and focused on the key requirement of adopting a process to make change happen. They also identified the importance of developing that skill in others below them.[1] (Graen, in his chapter suggests new methods to manage these leadership tasks.)

No matter the industry or economic environment, effective leaders lead others to achieve adaptability as a way of life. This means the organization and its people continuously find and define important problems, solve them, and implement valuable solutions. These problems may be strategic—defining a new vision or mission, establishing high-level goals, finding new directions to pursue or exploring new markets to enter—or they may be tactical, such as finding, solving and implementing opportunities for new products and technologies or for speeding up procedures. Effective leaders lead others to proactively sense, surface, discover, identify and define such problems and push toward implementation of solutions. By doing this continuously and by involving others, they lead their organizations or teams to make adaptability a way of life.

In order to lead people through this process in a synchronized fashion, leaders must learn to become process leaders rather than subject matter experts (Basadur, 2004). Simply defined, content is what we do, and process is how we do it. To leverage the thinking skills of employees, leaders need to engage them in the process of learning to think innovatively, rather than telling them what to do. This is called leading by being "the guide on the side" instead of the "sage on the stage." (Grace's chapter suggests methods to achieve such peer coaching.)

When leaders focus on continuously finding and solving important problems, they concentrate on process. Leaders who focus all of their attention on content typically solve the wrong problems. Understanding the crucial difference between content and process allows leaders to involve others in the creative process in a way that will maximize the use of their content expertise and uncommon sense.

Building Capability by Understanding Preferences

As discussed earlier, Simplexity Thinking is a continuous circular creative problem solving process that begins with the deliberate seeking out (generating) of new problems and opportunities. The second stage of the process is conceptualizing, or formulating, defining, and constructing a newly generated problem. In the third stage, problem solving, evaluation and selection of solution ideas takes place, while the fourth stage results in solution implementation. The process then begins anew, as every implemented solution (action) results in the opportunity to discover (generate) new problems and opportunities.

While effective innovation requires strong performance in each of the four stages of the creativity process, our research has found that individuals, teams and organizations may prefer some stages of the creative process more than others. We call these preferences styles and suggest effective leaders must learn to synchronize the different creativity styles. In teams, for example, the members must learn to combine their individual preferences and skills in complementary ways. Our previous research found that teams composed of members with a diverse range of preferences performed better than teams made up of members with similar preferences, but had less enjoyment working together (Basadur & Head, 2001).

While most people enjoy some stages more than other stages, it is typical to see preferences that combine or blend styles. It is also common for people to prefer one style in particular, but also have secondary preferences for one or two adjacent styles. An individual's unique creative problem solving style blend shows only their preferred activities within the creative process. These activities are illustrated in Figure 5.5.

No single style is to be considered any more 'creative' than any other. Skills are needed to execute all stages. All four stages of the process require creativity of different kinds and contribute uniquely to the overall innovative process and innovative results. Successful leaders will recognize and communicate the key message that everyone has a different but equally valuable creative contribution to make to the innovation process. By allowing employees to capitalize on their preferred orientation, leaders can make work more satisfying, as well as pinpoint individual development opportunities.

By tapping resources in all four styles, leaders can also help a team or the organization to cycle skillfully through the full innovation process. Skilful synchronization of the preferred creative styles and activities (shown

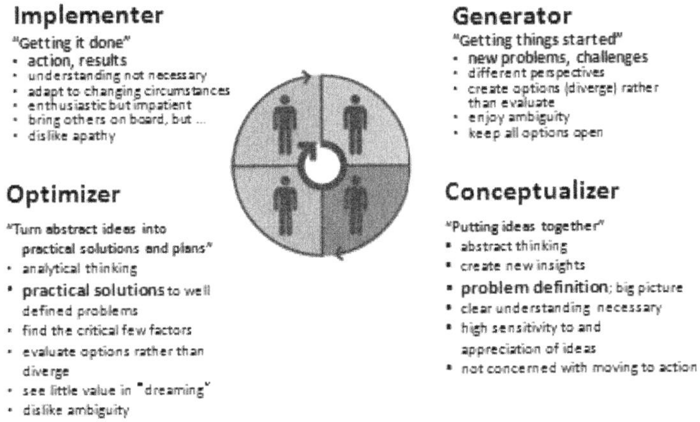

Implementer
"Getting it done"
- action, results
- understanding not necessary
- adapt to changing circumstances
- enthusiastic but impatient
- bring others on board, but ...
- dislike apathy

Generator
"Getting things started"
- new problems, challenges
- different perspectives
- create options (diverge) rather than evaluate
- enjoy ambiguity
- keep all options open

Optimizer
"Turn abstract ideas into practical solutions and plans"
- analytical thinking
- **practical solutions** to well defined problems
- find the critical few factors
- evaluate options rather than diverge
- see little value in "dreaming"
- dislike ambiguity

Conceptualizer
"Putting ideas together"
- abstract thinking
- create new insights
- **problem definition**; big picture
- clear understanding necessary
- high sensitivity to and appreciation of ideas
- not concerned with moving to action

Figure 5.5

in Figure 5.5) of interdepartmental and interdisciplinary team members is particularly important.

As Table 5.1 shows, the highest ranking Implementer style jobs include IT Operations, Customer Relations, Secretarial/Administrative Support, Project Manager, and Sales. From the handling of customer complaints to the need to minimize IT downtime, these positions all demand short term problem solving activities and quick delivery of results. The highest ranking Optimizer style jobs are Engineering/Engineering Design, Manufacturing Engineering, Finance, IT Systems Developer, and IT Programmer /Analyst. In each of these positions, practical, precise, and detail-oriented plans, processes and solutions are sought. The occupations that contain the five highest proportions of Conceptualizers are Organization Development, Strategic Planning, Market Research, Design, and Research and Development. These are all jobs in which understanding and problem definition are vital. Organizational, employee and customer needs must be defined so that new products, services, structures, and strategies for future growth can be designed.

Among the occupations that contain the highest proportions of Generators are Training, Marketing, Design, and Advertising. They require exploring new areas of inquiry; initiating new projects; seeking change and imagining possibilities for improvement, innovation and future growth. Marketing and Advertising are centered on initiating new projects and finding new ways to build interest among customers and capitalize on new trends and opportunities sensed in the environment. Designers initiate change by offering imaginative ways to communicate and stimulate interest in new ideas. Interestingly, people who describe themselves as designers

TABLE 5.1 Occupations Ranked by Occurrence of CPSP Style

Rank	Generators	Conceptualizers	Optimizers	Implementers
1	School Teacher	Organization Dev.	Engineering/Eng. Design	IT Operations
2	Academic	Strategic Planning	Mfg Engineering	Customer Relations
3	Artistic	Market Research	Finance	Secretarial/Admin
4	Non-Profit/ University Admin.	Design	IT Systems Developer	Project Mgr.
5	Training	R&D	IT Prog/Analyst	Sales
6	Marketing	Artistic	Accounting	Purchasing
7	Design	Product Dev.	Strategic Planning	Mfg. Prods.
8	Health Mgmt. Exec.	IT Sr. Consultant	Tech. Customer Support	Logistics
9	Advertising Mgr.	Academic	Social/Health Services	Operations

first prefer conceptualization (Stage 2) and secondarily generation (Stage 1). Other generator styles, including School Teacher, Academic, and Artist, are not prominent in industrial organizations. For these jobs, generator activities would suggest student development, music, art, writing, academic programs and research possibilities.

CONCLUSION

The successful leaders of the twenty-first century will be those who can coach their organizations and teams to make proactive adaptability a standard way of life (see Graen's chapter for example). This is more challenging than leading for efficiency, because it requires skills in coaching others to think innovatively—to continuously discover new disruptive problems and implement solutions. Getting people to think innovatively together requires a leader capable of protecting the divergent thinking of others. This includes building skills in being a coach—not simply a content expert—to help people move through the four different stages of the creative process. Mixing and matching different stages of the process and appreciating different ways of understanding things and utilizing such understanding is especially important in leading interdisciplinary or interfunctional teams, as various people in different kinds of jobs favor different stages of the creative process. Top leaders may also find organizational innovation and adaptability is strengthened when employees who prefer the generator style are recruited, motivated, rewarded and retained within their organizations (Graen, 2014).

For both organizations and individuals, successful and permanent adoption of an innovative mindset requires a permanent change in behavior, attitude and thinking. While organizational tools and techniques abound, most are well-intentioned concepts that are quickly cast aside. I predict that applied creativity—the process of finding, defining and solving important, complex problems, then implementing creative solutions—will result in the real and lasting organizational changes needed for success in today's business environment.

NOTES

1. Jeff Immelt, CEO of General Electric, initially made these observations during the CEO roundtable at the first Global Conference of the Procter and Gamble Alumni Association in 2003. His sentiments were echoed by others participating in the discussion. The importance of the point was immediately evident to me, and I've reiterated it in numerous papers (Basadur, 2004; Vincent, Stoyko, Henning & McCaughey, 2006) and presentations in the years since.

REFERENCES

Basadur, M. S. (1992). Managing creativity: A Japanese model. *Academy of Management Executive, 6*(2), 29–42.

Basadur, M. S. (1994). Managing the creative process in organizations. In M. J. Runco (Ed.), *Problem finding, problem solving, and creativity*. New York: Ablex.

Basadur, M. S. (1995). *The power of innovation*. London, UK: Pitman Professional Publishing.

Basadur, M. S. (2003). What is leadership in the 21st century? *CEO Roundtable, First Global Conference, Procter and Gamble Alumni Association*, Cincinnati, OH, April 26.

Basadur, M. S. (2004). Leading others to think innovatively together: Creative leadership. *Leadership Quarterly.* 15, 103–121.

Basadur, M. S. (2013). How might we? Three simple words that can drive economic prosperity in turbulent times. *The American University in Cairo Business Review, 1*(2), 82–87.

Basadur, M. S., & Basadur, T. M. (2011). Where are the generators? *Psychology of Aesthetics, Creativity and the Arts, 5*(1), 29–42.

Basadur, M. S., Basadur, T. M., & Licina, G. (2013). Simplexity thinking. Chapter in the *Encyclopedia of Creativity, Invention, Innovation, and Entrepreneurship. Springer*. Editor: Ruchika Bhatt

Basadur, M. S., & Gelade, G. (2003). Using the creative problem solving profile (CPSP) for diagnosing and solving real-world problems. *Emergence Journal of Complexity Issues in Organizations and Management, 5*(3), 22–47.

Basadur, M. S., & Gelade, G. (2006). The role of knowledge management in the innovation process. *Creativity and Innovation Management. 15*(1), 45–62.

Basadur, M. S., Gelade, G., & Basadur, T. M. (2013). Creative problem solving process styles, cognitive work demands and organizational adaptability. *Journal of Applied Behavioral Science, 50*(1), 78–113.

Basadur M. S., Graen, G. B., & Green, S. G. (1982). Training in creative problem solving: Effects on ideation and problem finding and solving in an industrial research organization. *Organizational Behavior and Human Performance, 20*, 41–70.

Basadur, M. S., & Head, M. (2001). Team performance and satisfaction: A link to cognitive style within a process framework. *Journal of Creative Behavior, 35*, 1–22.

Dolata, U. (2013). *The transformative capacity of new technologies: A theory of sociotechnical change*. London/New York: Routledge.

Gordon, W. J. J. (1956). Operational approach to creativity. *Harvard Business Review, 9*(1).

Gordon, W. J. J. (1971). *The metaphorical way*. Cambridge, Massachusetts: Porpoise Books.

Grace, M. (2014). Design, analytics and innovation: Welcome to the future of organizations. In M. Grace & G. Graen (Eds.), *Millennial management: designing the future of organizations. LMX leadership: The series* Vol IX. Charlotte, NC: Information Age Publishing.

Graen, G. B. (2014). Building a better boss: A proposed design thinking project. M. Grace & G. Graen (Eds.), *Millennial management: Designing the future of organizations. LMX leadership: The series* Vol IX. Charlotte, NC: Information Age Publishing.

Hazy, J. K., & Backström, T. (2014). Essential reins for guiding complex organizations. M. Grace & G. Graen (Eds.), *Millennial management: Designing the future of organizations. LMX leadership: The series* Vol IX. Charlotte, NC: Information Age Publishing.

Kabanoff, B., & Rossiter, J. R. (1994). Recent developments in applied creativity. *International Review of Industrial and Organizational Psychology, 9,* 283–324.

Mott, P. E. (1972). *The characteristics of effective organizations.* New York, NY: Harper and Row.

Short, J. C., Ketchen, Jr. D. J., Shook, C. L., &Ireland. D. R. (2010). The concept of "opportunity" in entrepreneurship research: Past accomplishments and future challenges. *Journal of Management, 36*(1), 40–65.

Simon, H. A. (1977). *The new science of management decisions.* Englewood Cliffs, NJ: Prentice-Hall.

Vincent, D., Stoyko P., Henning, G. K., & McCaughey, D. (2006). *Creativity at work: A leadership guide.* Ottawa, ON: CSPS Action-Research Roundtable on Creativity.

CHAPTER 6

LEVERAGE POINTS AND PROTOTYPES

Integrating Systems Thinking and Design Thinking to Help Organizations Evolve

Peter Coughlan
Colleen F. Ponto
Organization Systems Renewal Program
Seattle, WA

ABSTRACT

As organizations are confronted with ever-increasing rates of external change many are finding that their internal structures (and resulting behaviors) no longer allow them to respond effectively to that change. This chapter explores the integration of two theoretical approaches frequently employed to help organizations deal with change—systems thinking and design thinking. It argues that systems thinking and design thinking are required leadership skills as the rate of external change continues to accelerate and leaders must help their organizations adapt quickly in the face of this continuous change. In the course of individually applying systems thinking (Colleen) and design thinking (Peter) as core components of organizational systems change, the two authors share their experience of bringing together these two approach-

Millennial Spring: Designing the Future of Organizations, pages 107–122
Copyright © 2014 by Information Age Publishing
All rights of reproduction in any form reserved.

es in order to help client organizations identify potential high-impact leverage points for change and create novel responses that enable an organization to move in positive new directions. Through a series of workshops and client engagements, the authors have experimented with various ways to bring systems thinking and design thinking together in service of helping organizations create and implement a future of their own design. This chapter first looks at the strengths and limitations of systems thinking and design thinking and then shows how the integrated model has evolved based on its application to a variety of organizational challenges. The resulting model suggests a low-cost, robust, repeatable process for: (1) identifying where an organization might choose to intervene in order to have the greatest chance of impact; (2) generating potential future states; and (3) introducing those potential future states into the system in order to begin to shift the organization in a positive direction. The chapter concludes with a summary of what the authors have learned along the way in their attempts to integrate these two approaches and their future plans.

INTRODUCTION

As organizations are confronted with ever-increasing rates of external change (for instance, pressures from heightened global competition, the introduction of industry-disrupting technologies, changing demographics and consumer needs, or shifts in the regulatory or legislative environment, to name a few), many organizations are finding that their internal structures (and resulting behaviors) no longer allow them to respond effectively to that change. This chapter explores the integration of two theoretical and applied approaches—systems thinking and design thinking—that have until now usually been employed in isolation from each other to help organizations more effectively respond to both the external and internal changes they face. The authors propose that the combination of these two approaches, as well as the core capabilities of systems thinking and design thinking, will be central to leadership of twenty-first century organizations—organizations which are undergoing change at such a rate that they obsolete the traditional management skills taught to leaders of the more stable, slow-to-change organizations of the past century.

In the course of individually applying systems thinking (Colleen) and design thinking (Peter) as core methods of organizational change, the two authors share their experience of bringing together these two approaches in order to help client organizations identify potential high-impact leverage points for change and create novel responses that enable an organization to move in a positive new direction. Through workshops and client engagements the authors have experimented with various ways to bring together systems thinking and design thinking in service of helping organizations create and implement futures of their own design.

The chapter first looks at the strengths and limitations of systems thinking and design thinking and then shows how the authors' integrated model has evolved based on its application to a variety of organizational challenges. The resulting model suggests an efficient, robust, and repeatable process for (1) identifying where an organization might choose to intervene in order to have the greatest chance of impact; (2) generating potential future states; and (3) introducing those potential future states into the system in order to begin to intentionally shift the organization in a new direction. Bumps along the way in the authors' attempts to integrate these two models reveal important aspects about organization design and change and how we might help equip organizations and leadership with the tools, skills, and processes needed to continuously evolve themselves in response to changing internal and external conditions. The authors hope that by sharing the evolutionary journey of the model they can encourage further dialogue, experimentation, and education among a broader group of theorists, practitioners, and educators in order to speed development and adoption of what they believe has potential as a powerful and scalable approach to organizational change.

TWO POWERFUL APPROACHES
TO ORGANIZATIONAL CHANGE

Before beginning to work together in 2009, each of the authors had built careers based on a specific theoretical approach to organizational change. For Colleen, her primary approach to organizational change—systems thinking—began early in her career as a process engineer in a pulp mill where she used computer simulation models, complete with hundreds of feedback loops, to predict how changing variables such as temperature, pressure, and chemical flow rates would impact product quality. This early training in system dynamics, although not called system dynamics or systems thinking at the time, helped fuel Colleen's affinity for and grasp of systems thinking when she was formally introduced to the field several years later in graduate school, in the Organization Systems Renewal graduate program.

Having experienced the power of systems thinking and its potential for supporting personal and organizational transformation, Colleen came to regard systems thinking as an essential leadership skill of our time in order to effectively solve complex challenges and realize opportunities in human systems. Supported by this new mental model, she set out to increase her mastery of systems thinking and she focused her teaching and consulting career on systems thinking skill development. In helping individuals and organizational teams address current undesirable trends using systems thinking, Colleen repeatedly witnessed the power of systems thinking as an approach and set of tools useful in realizing the preferred future by first better understanding the current state of the system.

Peter's first exposure to his primary approach—design thinking—began during graduate school, when he had the opportunity to work on design projects that also happened to be highly systemic in nature—re-conceiving the kitchen systems for a fast-food restaurant chain, and designing a technology company's next-generation delivery platform for a new cellular communication standard. Both of these multi-month projects required their design teams to balance high-level conceptual views of complex, integrated systems with tangible, detailed design of specific user moments or touch-points. The work revealed the power of design tools to render complex systems visible and understandable; to help designers coordinate, through the creation of "user experiences," multiple structural elements that comprised a larger system; and through prototyping, to help a client organization begin to understand how newly proposed structures might begin to shift an existing system in a desired new direction.

From this early exposure to design challenges of a truly systemic nature, Peter was able to begin to understand the role that organizational and market context played in the launch of new ideas. He began to explore application of the design process to help client organizations deal with the organizational change required to introduce new products and services to the world (for an early articulation of this approach see Coughlan & Prokopoff, 2004). He discovered that enabling clients to author their own new ideas (rather than giving over idea generation to outside design consultants) often resulted in concepts that were better aligned to current organizational capabilities and that were more likely to be accepted by the organization since they were "invented here." He also discovered that introducing new product and service experiences into the system as "design experiments" enabled the organization to test and refine those ideas in low risk ways while simultaneously challenging the existing system to begin to behave in new ways. In other words, the use of a design process to facilitate change appeared to reduce initial organizational resistance while at the same time increasing the likelihood that proposed changes would be adopted by an organization.

THE INTEGRATION OF TWO APPROACHES—OUR HISTORY OF EXPERIMENTATION

In spite of the power we found in each of our respective methodologies, we both had our suspicions that something more could be done to strengthen our different approaches. For Colleen, there was a belief that while systems thinking was a powerful way to thoroughly understand the current state of a system in order to most successfully intervene in the system, the tools of systems thinking made it difficult to move from the present state to a future state.

Peter believed that while design thinking enabled teams to consistently generate elegant and inspiring future state scenarios, it was difficult to know—beyond reliance on a design team's intuition—where specifically to focus one's resources in order to create sufficient momentum and sustainable change for an organization. In spite of the power of design to help organizations generate novel products, services, processes, and experiences, Peter was still not confident in the ability of design tools to identify the best places to intervene in a system. Newly emerging organizational systems change models at the time such as Scharmer's (2009) Theory U or Cooperrider's Appreciative Inquiry (Cooperrider, Whitney, & Stavros, 2008) were beginning to incorporate design-inspired steps such as ideation and prototyping—further evidence of the benefit of design-based tools and methods to organizational change efforts—and yet it was not yet clear how systems thinking could contribute to the design of systemic change in organizations.

Shortly after his introduction to systems thinking through his affiliation with the Organization Systems Renewal graduate program in Seattle, Peter invited a group of IDEO designers to a systems thinking conference to explore whether they thought systems thinking approaches had anything to offer design thinking. His team concluded that while the field of systems thinking did indeed appear to have additional useful tools that could supplement the design process, there didn't appear—from their brief exposure to systems thinking at the conference, at least—to be a robust systems thinking process that could be used as a starting point for a combined systems thinking / design thinking approach. When Peter shared this conclusion with Colleen, she committed to familiarizing Peter with a process that had been developed and refined over the better part of a decade to introduce systems thinking to clients and to structure their organizational change programs (Goodman, Karash, Lannon, O'Rielly, & Seville, 1997; Kim, 1992; 1999; Meadows, 2008; Senge, 2006; Sweeney, 2001; Ponto & Linder, 2011). At that point, Colleen and Peter each became intrigued about the possibilities that their respective methodologies could offer the other. Inspired by their early collaboration, Colleen was quick to see the potential benefits of integrating design thinking and systems thinking and began right away to integrate design thinking into the approach she was teaching and using at the time. It was clear to her that design thinking could help move insights gained from a systems analysis into action (something that was lacking from a straight systems-thinking approach) and increase the likelihood of realizing the long-term benefits of an organizational change effort. Figure 6.1

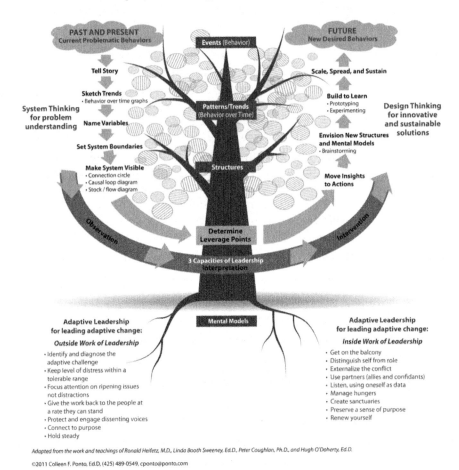

Figure 6.1 Our first attempt to represent the integration of systems thinking and design thinking.

shows how she initially incorporated key aspects of design thinking into the systems thinking process.

In this model, the tree at the center represents 4 levels of a system inquiry, starting at the top of the tree with the most tangible and visible level (events) and moving to the most intangible and invisible (mental models) represented at the base of the tree. Past and current problematic behaviors are revealed by following the first 5 steps of the systems analysis process—tell the story, sketch trends, name variables, set system boundaries, and make the system visible—as seen on the left side of Figure 6.1. Each step reveals more of the intangible and invisible aspects of the system being studied, such that once the system is made visible and causal connections among key system variables are mapped, areas of greatest leverage become obvious. At this step, leverage points can be identified; and with leverage points identified, new structures and mental models can be proposed, prototyped, tested, and launched. The model also incorporates the 3 capacities of leadership taught in the Organization Systems Renewal program and these three capacities do well to describe the high-level progression of activities—from observation of the system to interpretation of insights to intentionally designed interventions (which have been informed by a thorough systems thinking process).

In hindsight, this model works very well and is very close to the process that we currently espouse. Peter was reluctant to embrace the model entirely, however. He reasoned that if designers go through a similar process of observation and analysis (through such activities as telling stories, looking for patterns, and creating frameworks that visualize the dynamics in a system) then he was afraid that the handoff of a systems analysis to a design team would lack the richness of detail, personal engagement, and analogous thinking (inspiration and ideas taken from other similar experiences) that were important to any successful design team's work. Furthermore, if the design process relies on intuition to reveal unseen or even unconscious aspects of a human system, then would a rigorous analysis of the current state, involving behavior-over-time graphs and causal loop diagrams prevent the intuitive leaps that are the hallmark of great design from happening?

When the opportunity came to design and lead a workshop that would allow us to test a combined systems thinking and design thinking process, the authors used the chance to create a new iteration of their systems thinking/design thinking integration (Coughlan & Ponto, 2011). In this second iteration, the process would not just be a straightforward combination of systems thinking followed by design thinking, but rather a process that more tightly wove together the two approaches from beginning to end. The belief driving this design was that aspects of both systems thinking and design thinking might be relevant throughout an integrated process that moved a project team from understanding the current state of a system to envisioning and implementing future states of that system. For example, just as systems thinkers start their process by telling a story to set the context for a subsequent systems analysis, design thinkers frequently begin their process by telling stories to reveal aspects of the context that will provide inspiration for later design work. Could there be value in combining the different types of stories that systems thinkers and designers told at the beginning of each of their respective processes? Similarly, systems thinkers look for behavioral patterns and mental models to reveal underlying system dynamics, and designers seek to understand behavioral patterns and mental models in order to reveal opportunity areas that become the basis for future state design. Would the outcomes be better if a team looked for patterns aided by a combination of systems thinking and design thinking tools? Furthermore, could a process in which systems thinkers and design thinkers worked side by side mitigate the potential fears that the analytical approach of systems thinking would prevent the intuitive, creative approach of design thinking from working? With these questions and beliefs to guide us, we designed and led a workshop in which we more tightly wove together our processes—alternating between systems thinking and design thinking throughout the duration of the workshop. The resulting model is shown in Figure 6.2.

In this model, we clustered the combined process steps into 4 major blocks—define the challenge, ground our understanding, identify places to intervene, and move insights to action. Each of these 4 blocks is comprised of smaller process steps (which are similar to those found in Figure 6.1 above). The overall process we designed thus alternated between systems thinking (A) and design thinking (B) in an A-B-A-B pattern. Our intention behind this particular arrangement of steps was that the identification of leverage points in the system would be strengthened by having participants first grounded in their personal experiences through deliberate exploration of analogous contexts. In the same vein, design solutions would be strengthened because they were targeted to the most promising places to intervene in the system. We

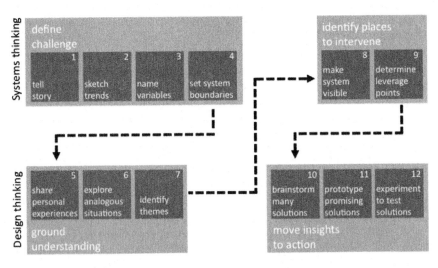

© 2011 Coughlan and Ponto

Figure 6.2   Our second attempt to represent the integration of systems thinking and design thinking.

thus tried to take what we thought were the best elements of both approaches and have each approach contribute to the other.

Our conclusions from this experiment with the integrated process were that (a) the shifts from systems thinking to design thinking throughout the process, and the sheer number of steps involved in the combined processes, were confusing to workshop participants, and therefore did not lead to an overall integration of the two processes; (b) creating the opportunity to insert design thinking earlier in the process did not appear to lead to more creative solutions; and (c) the sheer number of steps included in the combined process prevented participants from going very deep in any one step and caused an overall experience of feeling rushed throughout the process.

Following this workshop, we realized that a simplification of the process was in order and so we returned to Colleen's original hypothesis — that the best way to combine the processes was to first go through a systems thinking process to reveal the current state, followed by a design thinking process to generate possible future states, following a simple A-B pattern. This design is exemplified in Figure 6.3.

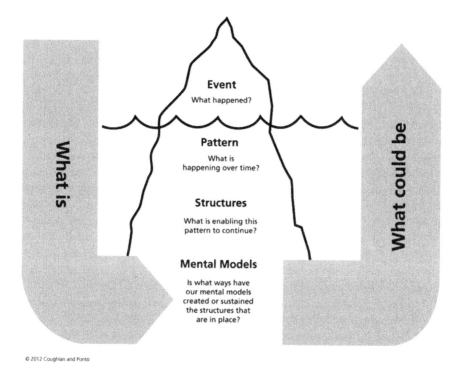

© 2012 Coughlan and Ponto

Figure 6.3   Our third attempt to represent the integration of systems thinking and design thinking.

This AB model has been described in our recent webinars (Coughlan & Ponto, 2012a, 2012b, 2013b) and has provided the structure of our three most recent workshops (Coughlan & Ponto, 2013a, 2013c, 2013d). In the first of these workshops, Colleen facilitated a client team through a day-long systems analysis before handing over an identified set of leverage points to Peter, who then ran a day-long design session with the client and dozens of external participants based on the pre-identified leverage points. In the second and third of these workshops, Colleen and Peter spent a half day performing a systems analysis to identify an area of focus followed by a half day of design work with a client team to generate potential future-state structures intended to be launched as experiments within the organization.

Overall, the model appears to be holding up in the context of real-world engagements—through the process, we are able to relatively quickly get a group of clients to identify recurring patterns they wish to change, to locate places to intervene in the system, to come up with possible structures that can shift the system, and to create strategies for introducing those structures into the system to begin to shift behaviors. We are pleased with the amount of progress that a client team and other, outside participants are able to make in a single day—through the process, we can efficiently move a group of up to 100 people from introducing one another to articulating the way in which interventions might be introduced into the client system.

A CASE STUDY

In order to give a sense for the nature of what gets done in a combined systems thinking / design thinking workshop, we describe the content of our most recent client engagement below (Coughlan & Ponto, 2013d). This workshop was conducted for the Annual Organization Systems Renewal (OSR) Alumni Conference, an event created to bring together alumni from the Master's Degree in Organizational Leadership for ongoing skill development and networking. In order to reflect the program's philosophy of "real work, real time, all the time," we reached out to the alumni to help us identify a possible client for our workshop. One alumna offered a challenge currently faced by the Early Learning and Child Care Division of the Boys and Girls Clubs of King County, Washington (BGCKC-ELCC). Prior to the workshop, we met with leadership of the organization and agreed on the following three objectives for the workshop: (1) to help client system (BGCKC-ELCC) increase understanding of organizational challenges they were facing in light of a recent reorganization (to centralize certain functions, including those housed within the Early Learning and Child Care Division); (2) to identify places to intervene in the system; and (3) to design specific interventions to be implemented as experiments by the client.

The day-long workshop, themed "Wisdom in the Room" because it is our belief that any system already contains sufficient wisdom to address its challenges, was attended by about 50 OSR alumni and 13 employees of BGCKC. We began the day by having the client give a description of the organization and tell us about their recent effort to centralize some functions that had previously been distributed among 40 Boys and Girls Clubs sites across King County. We then learned about the Early Learning and Child Care Division, its role in the larger organization, and some of their specific challenges, especially in light of the recent organizational changes. We next divided the room into different work groups, and distributed the BGCKC employees so that each work group had organizational representation.

We began our small group work with storytelling, asking everyone to tell a story related to the Boys and Girls Clubs challenge. Participants then identified possible variables at play in the system and created "behavior over time" graphs to represent recurring patterns or events in the organizational system. Each small group then selected one behavior to focus on moving forward. With that behavior as a focus, the group generated a list of possible structures or mental models that were keeping that behavior alive in the system.

Next, groups were asked to identify possible leverage points—places where an intervention could lead to possible behavior change. Some of the leverage points selected by the teams included:

- Local vs. regional control of resources
- Varied personnel policies and procedures across locations
- Impact of frequent organizational changes and initiatives
- Employees feeling heard
- Increase trust and accountability

We let each group select one leverage point that somehow resonated for the group. The selection process led to a range of different leverage points, which fit with our practice of addressing change by intervening in the system simultaneously through multiple leverage points. Teams next developed a series of "how might we" (HMW) questions from their identified leverage points. (These HMW's would serve as topics for the subsequent step of brainstorming.) Some examples of "how might we's" that groups created included:

1. HMW celebrate the uniqueness of each center?
2. HMW improve understanding of best/promising practices within and between sites and corporate?
3. HMW help sustain a culture where both children and staff flourish?

4. HMW increase trustful and authentic communication among all levels of the organization?
5. HMW reduce the pressure for directors to operate autonomously?
6. HMW enable communication between decision makers and implementers?
7. HMW help staff adapt to larger organizational changes while meeting site specific needs?
8. HMW balance local and central control?
9. HMW enable the system to connect and see the whole for optimization and innovation?
10. HMW engage employee voices for maximum impact?

Each team then spent 20 minutes generating ideas for new structures that could support shifts in behavior of the BGCKC employees. With the output from the brainstorming, teams selected two to three ideas to develop further through the creation of storyboards. Sample solutions developed through the storyboarding activity ranged from the adoption of existing products such as an anonymous mechanism to share data about local site performance with the executive team to new practices such as buddy systems or job swapping across sites to strengthen cross-site connection and knowledge sharing. Storyboards, together with possible experiments to introduce new structures into the organization, were presented to the clients, who were asked to provide immediate feedback. Some ideas clearly resonated more strongly than others, and it is these ideas that will likely be adopted by BGCKC as they seek to bring about change to their organization. At the time of this writing, the client reports that they are planning to run experiments from the ideas generated during the workshop, and there is considerable momentum around ideas that enable employees to spend time at different centers, thus addressing multiple leverage points with this approach. (This idea was also well received during the workshop.)

WHAT WE HAVE LEARNED

Perhaps the greatest discovery resulting from our experimentation is the utility of the iceberg model as the organizing framework (a backbone, of sorts) for the integration of systems thinking and design thinking approaches for organizational transformation. Working down the left side of the iceberg model with systems thinking helps clients clearly define the behavioral pattern or trend that they wish to shift (a problem to solve or an opportunity to realize) and helps them increase their understanding of the system through mapping and identifying the current structures and mental models in place. After leverage points are identified they are easily converted to "how might we" questions which serve as the foundation for the design thinking work that occurs as clients move up the right side of the iceberg model. Helping clients brainstorm, prototype, and implement new structures once the system is more deeply understood results in new behaviors and desired organizational trends.

Another key learning from our experimentation is around the variable of time—the amount of time we have to engage in the systems thinking + design thinking process with clients. As described above, we first experimented with two full days—one day to analyze the system using systems thinking processes and tools and a second day to create solutions using design thinking processes and tools. Even though this two-day engagement is probably ideal because we have time to create a detailed map of the current system, what we have learned in our recent one-day engagements is that in just eight hours we can help a client organization come to know their problem and system in a much deeper way, identify leverage points, choose places to intervene in the system, generate new solutions, and design experiments ready for implementation. Initial feedback from our two clients who participated in the one-day engagement model has been very positive.

A third key learning is that simple is good, even when dealing with analyzing and designing complex systems. Although we wanted to remain faithful to our respective methodologies and therefore not sacrifice steps from either process, the merging of two complimentary processes (with related steps) caused the process to feel unwieldy and unnecessarily complicated; and, as mentioned above, caused participants to feel rushed. We have learned that the respective contributions of systems thinking and design thinking each add value to the process of helping an organization with change—with each client engagement we are learning to trust the combined process as much as we have each trusted our own respective approach to organizational change.

FUTURE PLANS

We plan to continue to refine our combined approach through engagements with client organizations. We already have shifted the model based on our most recent engagements and evolving understanding of how systems thinking and design thinking complement each other. Ultimately, we will infuse each step in the unified process with systems thinking and design thinking—for instance, getting stakeholders to engage in contextual observation to better understand the current system dynamics. The newest iteration reduces the process to five steps—understand the system, make the system visible, determine leverage points/places to intervene, envision new possibilities, and implement promising solutions (see Figure 6.4).

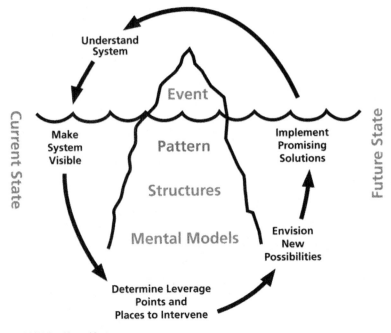

© 2014 Coughlan and Ponto

Figure 6.4 Our most recent attempt to represent the integration of systems thinking and design thinking.

This latest version of the model reflects the iterative nature of the organizational change process—introducing anything new into a living system shifts that system (no matter how intentionally designed), thereby creating a new current state with its own unique dynamics and possible interventions. It also more closely reflects our current understanding that mental models and structures provide valuable places to intervene in organizational systems, and that any process should help stakeholders to seek out these mental models and structures in order to create the most efficient and effective means to help organizations move constructively forward.

The model also reflects our own integration and mastery of the respective systems thinking and design thinking tool-sets. As we become more familiar with each other's methodologies, we are more and more able to suggest ways that a particular tool or method could be made more effective by applying a technique or a perspective from the other's toolkit. We imagine a day when teams will be equally familiar with systems thinking and design thinking, and the integration of each approach will be effortless and expected.

Results of our action research have shown to us and to our clients that the integration of systems thinking and design thinking is an effective approach to helping organizations solve tough challenges and realize exciting opportunities. Our experiences, coupled with the collection and analysis of short- and long-term feedback from clients will fuel the continued evolution of our process. Once we settle on a process that we feel is most effective and efficient we plan to scale and spread this social technology by teaching and coaching others to lead the process.

We are also excited to begin to research the impact that our methodology has on client systems. While we are of the opinion that a day-long intervention can cause a permanent shift in the way the client system views itself, we are excited to explore ways to increase the likelihood that experiments designed around leverage points will be introduced into the system. The introduction of experiments will require the development of structures and training to encourage ongoing experimentation with ideas conceived in the workshops, as well as training in how to evolve interventions based on feedback from the organizational system and how to manage a portfolio of potentially interrelated interventions. These, we believe, are the critical leadership skills of the next decade, and we are excited to be exploring them in our teaching and in application to our clients' organizational challenges.

From a pedagogical perspective, we have begun to think about how we can help deepen students' mastery of systems thinking and design thinking (and the effective combination of the two) so that they can both competently and confidently exercise leadership and address complex adaptive challenges in their organizations and communities. We believe that this blend of skills will play an increasingly important role in what it means to skillfully and successfully help organizations survive, adapt, and thrive in this ever-changing world.

REFERENCES

Cooperrider, D. L., Whitney, D., & Stavros, J. M. (2008). *Appreciative inquiry handbook for leaders of change* (2nd ed.). Brunswick, OH and San Francisco, CA: Crown Custom Publishing and Berrett Koehler Publishers.

Coughlan, P. J., & Ponto, C. F. (2011, November). *Systems thinking + design thinking for innovative and sustainable solutions.* Session presented at the Pegasus Systems Thinking in Action Conference, Seattle, WA.

Coughlan, P. J., & Ponto, C. F. (2012a, April). *Systems thinking + design thinking: Moving 'what was' and 'what is' to 'what could be.'* Webinar presented for Pegasus Communications, Inc.

Coughlan, P. J., & Ponto, C. F. (2012b, October). *Systems thinking and design thinking: An integrated approach to solving tough problems.* Workshop presented at the Relating Systems Thinking and Design Thinking Conference at the Oslo School of Architecture and Design, Oslo, Norway.

Coughlan, P. J., & Ponto, C. F. (2013a, March, April). *Real work, real time: Organizational design challenge.* Workshop presented at the April 2013 Organization Systems Renewal Alumni Association Annual Conference, Seattle, WA.

Coughlan, P. J., & Ponto, C. F. (2013b, September). *Systems thinking + design thinking: Designing organizational interventions that work.* Webinar presented for the Nexus4Change Scholar Perspectives Webinar Series sponsored by the Organization Development Education Association of the Organization Development Network.

Coughlan, P. J., & Ponto, C. F. (2013c, October). *Systems thinking + design thinking: Integrating two approaches to solve an organizational challenge.* Workshop presented at the Relating Systems Thinking and Design Thinking Seminar at the Oslo School of Architecture and Design, Oslo, Norway.

Coughlan, P. J., & Ponto, C. F. (2013d, December). *Wisdom in the room: Understanding what is, realizing what could be.* Workshop presented at the December 2013 Organization Systems Renewal Alumni Association Annual Conference, Seattle, WA.

Coughlan, P. J., & Prokopoff, I. (2004). Managing change, by design. In R. Boland & F. Collopy (Eds.), *Managing as designing* (pp. 188–192). Stanford, CA: Stanford Business Books.

Goodman, M., Karash, R., Lannon, C., O'Rielly, K. W., & Seville, D. (1997). *Designing a systems thinking intervention.* Waltham, MA: Pegasus Communications, Inc.

Kim, D. H. (1992). *Systems archetypes I: Diagnosing systemic issues and designing high-leverage interventions.* Waltham, MA: Pegasus Communications, Inc.

Kim, D. H. (1999). *Introduction to systems thinking.* Waltham, MA: Pegasus Communications, Inc.

Meadows, D. H. (2008). *Thinking in systems: A primer.* (D. Wright, Ed.). White River Junction, VT: Chelsea Green Publishing Company.

Ponto, C. F., & Linder, N. P. (2011). *Sustainable tomorrow: A teacher's guidebook for applying system thinking to environmental education curricula.* Olympia, WA: Pacific Education Institute.

Scharmer, C. O. (2009). *Theory U: Leading from the future as it emerges.* Cambridge, MA: Society for Organizational Learning.

Senge, P. M. (2006). *The fifth discipline: The art and practice of the learning organization* (2nd ed.). New York: Doubleday / Currency.

Sweeney, L. B. (2001). *When a butterfly sneezes: A guide for helping kids explore interconnections in our world through favorite stories.* Waltham, MA: Pegasus Communications, Inc.

CHAPTER 7

ESSENTIAL REINS FOR GUIDING COMPLEX ORGANIZATIONS[1]

James K. Hazy
Adelphi University
Mälardalen University

Tomas Backström
Mälardalen University

ABSTRACT

Informed by the complexity perspective, this chapter argues that effective leadership in today's complex organizations recognizes that structures which appear to be stable and predictable are in reality malleable and ever-changing. It takes a broad view of the design school model and considers leadership to be a design-in-action process that is enacted through others in response to uncertainty and changing circumstances. Complexity leadership continually enables the emergence of new structure by evolving and reinforcing recognizable and useful practices that can temporarily stabilize routine interactions. At the same time, it challenges and sometimes erases stale processes that resist or hinder change. To further specify this process, the chapter identifies three categorical dualities that frame how events are represented or described with-

Millennial Spring: Designing the Future of Organizations, pages 123–143
Copyright © 2014 by Information Age Publishing
All rights of reproduction in any form reserved.

in the workgroup. Effective leadership, it argues, influences the choices and actions of others by shaping the perception of events in the context of the six poles of these dualities. By doing so, it limits or expands what is perceived to be possible and at the same time defines and clarifies what is expected. The chapter describes how leaders tend the reins that guide collective performance along six essential complex organizing functions by alternatively tugging and loosening the constraints on thinking and action that are implied by the poles of each duality.

Everything existing in the universe is the fruit of chance and necessity.

—Democritus

In this chapter we argue that in today's complex organizations, leadership is even more difficult to define than in the past. This is because "leadership" exists within a complex mixing of unique personal preferences, idiosyncratic choices, evolving individual and collective purposes, and a widening web of social influence. Individuals and their leaders navigate this sea of rising and falling uncertainties as change and stasis interact and where predictability enables periods of comforting stability.

Effective leadership orchestrates the interactions between an individual's desire to act freely and creatively and with potency, and the constraining needs of the community that values orderly and controlled collective action (Barnard, 1938). It operates at the nexus wherein seemingly autonomous actions form integrated outcomes. This is why Mintzberg (1990) has criticized the design school: "In particular we reject the design school model where we believe strategy formulation must above all emphasize learning, notably in circumstances of considerable uncertainty and unpredictability, or one of complexity in which much power over strategy making has to be granted to a variety of actors deep inside the organization" (p. 190).

As such, we take a broad view of design school model in the tradition of Mintzberg (1990), and consider leadership itself as a kind of design-in-action process that operates at this nexus within complex organizations (Goldstein, Hazy, & Lichtenstein, 2010). While responding to uncertainty and changing circumstances, leadership acts to develop and facilitate a working-design to fit collective competences with emerging opportunities. By doing so, it enables the emergence of new coarse-grain properties and does this by evolving and reinforcing recognizable and useful practices that stabilize routine interactions (Hazy & Uhl-Bien, 2013a, 2013b). Katz and Kahn (1966) called this indescribable something "the influential increment" which also applies to complex systems (Hazy, 2011). This is relevant because it is at the same time both ethereal and deeply significant. Weber (1946) called its essence, "charisma" from the Greek meaning a gift "... believed to be supernatural, not accessible to everyone" (p. 245).

Most contemporary leadership research either seeks to reduce the essence of effective action within this nexus to some kind of interpersonal exchange (cf. Graen & Uhl-Bien, 1995; Bass, 1985, 1990), some learned behaviors or (Sashkin & Sashkin, 2002) or an emotional connection (Conger, 1989; Goleman, Boyatzis, & McKee, 2002) to be manipulated or feared. But, complex interaction cannot be reduced in this way.

We propose a different approach to exploring "leadership." In this chapter, we will look for the essence of leadership in the space that exists between the naïve belief that structures are stable, constant, and the contrary fretful paralysis that comes from the fear that everything is malleable, ever-changing and that one's only viable option is to hold onto someone who seems to know what is going on, the "leader," whether that primal trust is deserved or not.

To illustrate, we imagine two perspectives from a common scene in a genre of American "Western" films: In the mid-eighteenth century western United States, a runaway stage coach is being brought under control by the matinee hero, a handsome cowboy, in our case "the leader." This scene is typically filmed from two complementary perspectives. One, what we are calling the road map, is the perspective that pulls back the camera to define the structure of the problem for the audience—it "initiates structure." This is the cinematic wide-angle shot that follows a bumping, fast-moving and noisy stage coach stirring up a swirling cloud of dust against a backdrop of desert buttes and blowing sage brush. Sometimes the road is rocky; sometimes the coach must be guided along a narrow riverbank; at times the coach is under attack by competitors. This long view of the terrain gives the viewer a sense of the challenges facing the leader as he or she designs a way forward and how these designs might need to change with events.

The second perspective is the close-in shot. This is filmed from the point-of-view of the leader frantically, but confidently manning the reins to maintain just the right level of control for the situation, designing for others a way forward while being ready to adapt and change those designs as needed. This cinematic shot provides the emotional, personal perspective that is involved in relational considerations that are critical when guiding a team. The sustaining impression from the hero's point-of-view is that of wild motion, overwhelming animal energy and spirit, strength, exertion, resolve being tested, and of course, controlled confusion, until finally the team safely and triumphantly reaches its destination. As the situation is resolved, there is a rush of pride and relief, but also of confidence and a sense of power. This is what effective leadership feels like, or at least what it *should* feel like. Succeeding close-in requires detailed operating knowledge and an accomplished mastery of the design-in-action skill set. As such, it would be nice to have an operating manual to help master these skills. As we will describe in last next main section, to succeed in the context of the close-in-shot one must develop a specific set of skills with regards manning the reins of leadership (Backström, 2013).

The requisite tool-kits for succeeding at both the wide-shot shot and close-in shot are called the "roadmap" and the "operating manual" respectively. The former describes the functions performed by leadership in designing possible futures and the latter the dualities of meaning and of making sense (Weick, 1995) that can be used by leadership in the context of personal relationships to help the organization's members resolve their roles and understand their contributions.

Although informed through a complexity world view, these ideas are not new. To demonstrate this, we begin by reflecting back on the work of Barnard (1938) and Mintzberg (1981, 1990) who explored the interaction between individual choice and collective action that bounds our recognition of "leadership" and what it means to us in our organizational lives. This deeper understanding of experience is absolutely essential when guiding complex organizations over today's difficult business terrain.

THE REINS OF LEADERSHIP

Barnard (1938) described a key duality within organizations, the perspective of the individual's choice or *autonomy* versus its position within a social system, or *integration*. "These two aspects are not alternative in time.... Rather they are alternative aspects which may simultaneously be present. *Both are always present in cooperative systems.* The selection of one or the other of these aspects is determined by the field of inquire" (pp. 16–17). These are the reins of autonomy and integration.

We will show that this duality and the choice of inquiry are complemented by a second duality which reflects the information conditions in the environment. In particular, there can be conditions of *divergence* when new information must be learned from emerging structures, and there can be conditions of *convergence* toward current competencies when nonconforming information is often discarded and forgotten. Both can be present at the same time even in the same event. As Mintzberg has written, "Every strategic change involves some new experience, a step into the unknown, the taking of some kind of risk. Therefore, no organization can ever be sure in advance whether an established competence will prove to be a strength or a weakness" (p. 182). These are the reins of divergence and convergence.

Finally, we describe how this relates to leadership theory as epitomized by a third duality. This third duality highlights an individual's potential to enact meaningful outcomes, to realize one's own *potency* within the larger social system. These individual possibilities are counterbalanced by the impeding force of institutional *constraints* acting on this potential through social (as well as physical) contexts. If one ignores these important dynamics, "… the attention transfers to the organization and as a whole, or to remote parts of it, or to the integration of efforts accomplished by coordination, or to persons regarded in groups, then the individual loses his preeminence in the situation and something else, non-personal is character, is treated as dominant" (Barnard, 1938, p. 9). These are the reins of potency and constraint.

These dualities are shown in Figure 7.1. The right side of the diagram illustrates the complex organizing environment confronting today's leaders and managers. On the one hand, new opportunities and threats arise and change quickly, and these feedback on one another to create entirely new challenges. In these situations, individuals cannot rely on the "company line." Instead, although they might be tempted to look to their left and then to their right for guidance, eventually they realize that they are left to their own devices and must act with autonomy (the top left of Figure 7.1). This results in many autonomous actions that reflect divergence in the information being created in the environment. On the other hand, as these new structures emerge and are reinforced, they become institutionalized and exert pressure on individuals to conform. New information is discarded or ignored. There is convergence toward "the way things are done around here."

By influencing what others experience and what they believe about themselves and the situation through the thoughtful use of these dualities,

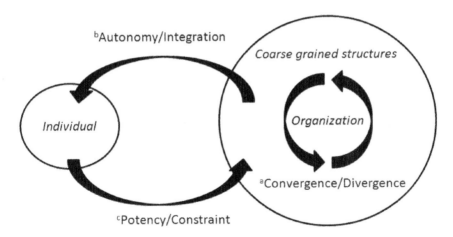

Figure 7.1   The three dualities: a. Convergence|Divergence has to do with the development of coarse grained structures at the organizational level. b. Autonomy|Integration is about the resources and information available to the individual to stand free from or follow the influence of coarse-grained structures. c. Potency|Constraint is the individual choice to stand free from or follow coarse-grained structures.

individuals, what might be called "leaders" can actualize their own level of potency (the bottom left of Figure 7.1). At the same time, all individuals must learn to circumvent, overcome or succumb to the constraints acting upon them from the various institutional pressures that seek to limit their capacity to make a difference.

Although each duality above is named in terms of its poles, for each, what is relevant in a given situation lies between these poles. This is why we call the poles of each duality "the reins of leadership." When a duality is skillfully reined, one can use the reins of leadership to unlock the inner power of collective action. However, it is important to note that these reins neither control us, nor (absent coercion) do they allow us to control others. They are but a connection, a connection that must be countenanced wisely and skillfully. This is the central challenge of leadership as one seeks to navigate the difficult terrains of organizational life. When knowledge of these dualities and what they mean is thoughtfully applied in communication and in human interaction, they represent a means of influence over organizational outcomes, and to a degree even of control

LEADERSHIP MUST PERFORM SIX ORGANIZING FUNCTIONS—THE ROAD MAP

This section provides a road map for leaders and managers to help them design and evolve their leadership approaches to succeed across changing business terrains. It does this by clarifying the various functions that leadership—and thus the leader—is called upon to do. Having a clear road map of the functional demands of leadership will help the organization's members identify where they are, determine what obstacles they are facing from a business perspective, and visualize where they need to go.

More specifically, this section provides a roadmap that helps one recognize each of six distinct terrain types—what we call *functional demands* that leadership must navigate when organizing collective action—and it explains the characteristics of each and their potential pitfalls. Furthermore, it describes how and why each terrain must be traversed differently using unique skills and calling upon different experiences. Taken together, the six terrains facing today's organizations constitute the elements of the roadmap that will enable managers and leaders to design performance excellence into their organizations as they lead. They do this while enacting these six critical (and difficult) functions:

1. Recruiting, selecting, retaining, training, and motivating high quality participants in support of the community.

2. Gathering and synthesizing relevant intelligence distributed across the organization, customers and other stakeholders,
3. Enacting community-building to unify individuals into effective work groups,
4. Generating options through initiatives and performing experiments to test hypotheses about the viability of future possibilities,
5. Structuring activities, resources and people by using available information to select and address opportunities, and
6. Executing plans and programs that sustain the organization by initiating activities such as task assignment, monitoring, control and feedback on outcomes.

No organization today can survive for long without leadership competence that guides people through these six terrains toward a meaningful *end*.

Recruiting and Motivating Participants (RMP)

Individual agents must be brought into the community and socialized into its norms. For the community, individuals must be recruited, motivated and retained, and in some cases, individuals must be disciplined or removed. This is a standard terrain for leadership—developing people and motivating them in the context of the collective purpose being espoused. As such, we say little more on this function that has been well-studied in social psychology.

Gathering and Synthesizing Information and Intelligence (GI)

In the ancient text, *The Art of War* Sun Tzu famously said, "...what enables the wise sovereign and the good general to strike and conquer, and achieve things beyond the reach of ordinary men, is foreknowledge" (13:4). This has been a tenet of effective command leadership for centuries. Surprisingly, little in modern leadership research addresses the challenge of leading an organization in the acquisition and synthesis of intelligence about the marketplace and even the functioning of the organization itself (also critical). It is as if this part of leadership must be kept secret.

This misperception feeds the mysticism. Leaders are expected to know things; they know what is happening, they know what you are thinking; and they know just what to do when everyone else is fiddling about. However, a moment of reflection allows one to realize that this perspective simply cannot be true. The leader may be ahead in terms of knowing what happens, but this doesn't happen by accident nor by "charisma."

One of the first stages in leadership is getting good information, making certain it is unbiased, and continually insuring it is up-to-date. As Sun Tzu tells us, this will ensure victory. As such, this organizing function, like the others, must be led. This organizing-terrain that must be recognized and crossed by leadership is called *gathering and synthesizing information and intelligence.*

Community-Building (CB)

To succeed in organizations, sometimes individuals must subordinate their individual drives to the needs of community. This cannot be done as plans are being executed. Rather, the seeds of community-identity must be sown before action ensues, even though the needs of the community must be clarified and reinforced during decision and action. This is why community building is a separate terrain for leadership. It unites otherwise autonomous individuals into a common unified entity, we call this the *community-identity.*

In any organizational setting, individuals are both autonomous and integrated in some way; as people, they each desire individual expression, or potency, but also a sense of belonging even though to belong brings with it a certain level of constraint. Norms and customs, perhaps even rituals, are followed; common language is used; legal or other institutional strictures are enforced; perhaps one can even be banished or shunned for not following along with the constraints. Often, in an effort to further convergence towards some objective, divergent thinking is ostracized or ignored discarding much of the information already present in the system. How and why this occurs and is useful represents the community-building function, yet another terrain that is essential to for leadership to traverse successful.

Generating Options (G)

The fourth leadership terrain that leaders must first recognize and then effectively traverse involves adapting to change. More specifically, leaders must have the capacity to enable the organization to develop options or in evolutionary terms "variations" that can be tested in the environment, experiments in an organizational context. This capacity can most simply be summarized as trying things to see what works. Of course, this is a gross oversimplification because one must experiment thoughtfully to get the most benefit from such activity. Otherwise, it is easy to waste time and resources with nothing to show for it. This requisite organizing function for which leadership is responsible is called *generating options* or possibilities.

In the business context, options often have inherent value, including financial value called *real option value,* that comes from a positive expected value related to broad variance but with limited downside risk. By thoughtfully truncating the downside of the probability distribution while maintaining the long tail of the upside, the organization is able to achieve a positive expected value even on very low probability projects. A critical prerequisite to achieving real option value, however, is the thoughtful construction of the experiments such that by their structure, they create rather than dissipate value. And these thoughtful choices must be taken *before* there is a clear picture of which experiments might be adaptive and which will lead to dead ends. This navigation through uncertainty is a function of leadership. It is the difficult terrain often referred to as leading innovation or leading emergence. This terrain too must be recognized and crossed successfully for long term sustainability to be achieved.

Structuring Activities by Using Information (UI)

This leadership function unfolds in the context of the information that has been gathered by individuals, groups and organizations and is for the most part distributed among them, unavailable explicitly to others even the leader. This situation creates a unique, under-researched leadership challenge: how does one make implicit the diverse and disparate information hidden in your people, information and personal knowledge that has been hard won by them through their own unique experiences.

This private information can become explicit through the design of organizational experiments. Although in practice design is often considered to be an "end" in itself, it is really a means to find hidden private information that might be relevant to the broader community. This fact comes to light as the group test structures that have been proposed according to some purpose. This community-wide shared information continues to flow forth during the process of continuous iterative trial, feedback, refinement or elaboration, further trial, feedback and so on and so forth.

Thus, design for feedback is critical. This happens as those who construct the experiments make assumptions about what will happen, try it out, and then *collectively recognize* the information created by the experiment. In an experimental context, all those involved recognize the information because it is created in the context of the experimental design which they share. The disparate information that had been stored within individuals' memories becomes manifest in the emergent structures that survive as the experiments are designed and executed, and the feedback is processed into the next iteration.

As information is gathered, reference models are created and shared, more information created through experiments, and this is further gathered and used to improve models to guide further experiments and so on, as activities are structured by using information (Surie & Hazy, 2006). Similarly, the old, discredited ways of doing things must be erased to prevent backsliding. Many organizations fail in the leadership of this critical function allowing ineffective or counterproductive aspects of organizing to remain in place when in fact they are harmful. This leadership terrain involves recognizing the need to build useful structures into organizational forms, and enabling this process. At the same time, it also recognizes that the organization must also thoughtfully "erase" the information that draws activities in unproductive directions (Hazy, 2012). The failed approaches or stale old ways of doing things must be discarded and banished. Leadership has a role in ensuring that bad organizational habits are eliminated, just as it is responsible to grow new ones.

Executing Plans and Programs (E)

Independent of structuring activities it is a function of leadership to execute the structures effectively once they are adopted, to improve them and to continually drive efficiency into the process. This terrain can be recognized when there is a clear objective and when the process for achieving the objective is likewise well defined. Under these conditions, distractions and disagreements impede progress rather than further it. Clarity of roles, integration of activities, individual competence at specialties, and clean procedural handoffs are predictors of success.

The analogy here is of the sports team which runs plays to score its goals. How effectively the play is executed is just as important, perhaps even more important, than the skills of the individuals players. This is an active function of leadership: to run the plays effectively. Individuals who seek to lead must learn to recognize this terrain because on this terrain, divergence, autonomy and even a drive of individual potency (sometimes called "hotdogging" in sports) can hurt execution.

DIFFERENT DUALITY-REINS FOR VARYING TERRAINS— THE OPERATING MANUAL

The reins of connection between the self and the system can be used to achieve safe passage across the six different functional terrains described in the roadmap. They provide the *means* at the leader's disposal to navigate the organizational landscape. They enable him or her to recognize and to make sense of complex organizing situations. Achieving success however requires the skillful use of the reins of leadership in the context of a deep understanding of these categorical dualities:

- divergence | convergence
- autonomy | integration, and
- potency | constraint

One must go beneath the obvious to understand how an organization can *be* both aspects of each of these dualities at once as shown in Figure 7.1. Indeed, what is happening can only be understood when one realizes that complex organizing requires *being dynamically engaged* with all six of these reins—the end-point categories for each duality—at once, and with everyone in the organization at once.

As the reader might have anticipated, each of these six end-point categories also represents the primary rein to be mastered when driving the organization over one of the six functional terrains that must be traversed by leadership. The categorical dualities reflect the challenges of effective leadership and provide a hint as to why leadership is so difficult to describe. As such we divide the operating manual into six sections, each describing how to navigate a different organizational terrain.

In the sections that follow, although for each terrain type we identify a single rein as the focus, we want to emphasize that all six reins must always be thoughtfully managed. Only by attention to the entire system all of the time is it possible to maintain the proper feel for the organization's situation, direction and potentials as terrain types change quickly and unexpectedly. On Monday one might need to lead the organization in gathering intelligence, but by Wednesday, one might need to recruit and train a lieutenant to replace a key player who has been reassigned.

This philosophy of leadership has to do with the dynamics at work within organizations, the stability and change. The approach has been used to design training for leaders in a research environment called the Innovationsgym at Mälardalen University in Sweden. The training program is designed as a series of workshops that we have found can be used to address each terrain type and rein (duality pole). The workshops are given in the order described below to reflect what we believe is a common unfolding of leadership challenges in organizational life Examples from this training will be given at the end of the description of each of the six parts of the operating manual. In practice, however, the reader should note that there is no natural ordering. All reins are possible because all terrains are possible. And they can change suddenly and in unexpected ways.

Recruiting and Motivating Participants by Tugging
Individual Potency

How does one motivate people to accept new and potentially unknown structures that might come out of a change process? How does a leader help others abandon their old comfortable way of doing things? These are common questions in the organizational change literature. One standard answer is to let people participate in the change process. The thought behind this management philosophy is often described as using the resources of the employees to improve the change process itself. Often, however, "gaining the employees buy-in" is also given as a benefit of this approach. We think of it this way: one can motivate people to accept new approaches and to abandon old ways of doing things by awakening in them confidence and a willingness to go in a new direction by tugging on the rein that increases their *individual potency* and their sense of self-reliance and self-efficacy.

The importance of individual potency can be seen, for example, by looking at cases where companies survive though innovations that grow from "skunk works." This is where a group of people develop radical innovation in an environment where they are shielded from the coarse grained structures of the organization. Sometimes skunk works are established and supported against the direct orders from managers who have assessed them as unprofitable. Giving people a clear permission to let go of old structures, unrealistic instructions, common routines and orders may be a way to enhance individual potency. Some Japanese organizations use a red feather at the door post as a sign when there is a meeting where you are supposed to use your individual potency and ignore the coarse grained structures of the work place.

It is common in organizations to insure there is a high degree of diversity among the people included in the innovation process. This is done to increase the chance that many new ideas covering a large part of the space of all possible solutions or opportunities will be voiced. However, there is still a need to empower individuals to use their voices. This is done by tugging the rein connecting the leader to each individual's sense of personal potency and involvement making it clear that such participation is expected, wanted and valued.

There are other interventions that can enhance individual potency and enable others to engage in radical new thinking. Human beings can themselves be stuck in their own everyday structures, their habits of mind and doing. In the workshops in the Innovationsgym we use tools like meditation, perspective shifts and safaris at one's own workplace to cleanse people of their old structures and give them a sense of the possibilities that await by thinking in new ways. One can also be trained to consciously look for new starting points in one's thinking. For example, one can open-mindedly gather new intelligence or engage the perspectives of others—new and potentially useful information—by meeting with customers, colleagues, etc. These interactions have special power when one's mind is empowered by the possibilities opened up through an awakened sense of individual potency.

Gathering Intelligence by Tugging *Divergence*

As was observed, in the ancient text, *The Art of War* Sun Tzu famously said, "...what enables the wise sovereign and the good general to strike and conquer, and achieve things beyond the reach of ordinary men, is foreknowledge" (13:4). In this section we describe how this insight is relevant to leadership, and also how this function of leadership is enacted by tightening and loosening the tension in the reins of leadership, with particular attention to one of those reins: *divergence*.

Gathering and synthesizing information and intelligence is related to the level of divergence within organizations for several reasons. First, there is a lot of information out there due to the continuous flow of events. Just recognizing the information created through events requires a certain expectation about what is likely to happen. Absent some investment in sensing devices to *observe* the event, what actually happened will not even be noticed. Second, the process of selecting on which observed information to focus attention is likewise an important one. The judgment on what information might ultimately be usable can seldom be grounded on standard organizational procedures. It must be based on the knowledge and skill of individuals as informed by those individuals' specific integration into their social structures. Third, there is also normally a need for a translation process: how does this external information relate to other information; what does it mean for us in our situation? How can we make use of it?

Recognition, selection and *translation* are all processes where different individuals notice, decide and act differently. All people taking part in the intelligence gathering process will observe new information, but because of each person's unique position, each will observe different new information, select what might be considered relevant information differently, and also develop a different understanding of it. The gathering and synthesizing information and intelligence is thus directly related to the level of divergence of perspectives within the organization. Relaxing the divergence rein might cause the organization to miss something important. Tugging the divergence rein too tightly might allow the too much confusion to enter the system, threatening its unity. Effective leadership navigates this terrain.

Exploration after new knowledge and ideas in the external world is part of the first and fuzzy end of a development process that is an area of focus in the Innovationsgym workshops. Raw and unsorted information in the external world is collected through divergence. When integrated into the prior perspectives this new information becomes part of the tacit knowledge shared among the people included in the process. Pictures and photos are often used in this part of the creative processes to keep the openness of the information and to identify associations in the interactions before trying to precisely define and close the information to something more specific.

Unifying Individuals into Work Groups by Tugging *Convergence*

A lot of groups in work life are really not more than some individuals working in parallel, hopefully towards the same goal. For them to be able to use each other's competences, to have synergy affects and to discover new fields between their different types of expertise, they have to be unified into a community, a unified group. Coarse-grained structures will emerge during this process of getting-to-know and coming-to-trust each other as each learns the nuances of the other's knowledge, appreciate the other's perspective and interprets all of this in relation to the common task. Convergence always results from this emergence of common understanding.

There are different ways to manage projects in work life. Most commonly a plan and control strategy is used, for example a stage-gate model. But there are also examples of a more emergent strategy where each project member is given a picture of the total project, where responsibilities and resources for real-time handling of problems are distributed to all members, and where control is gained by interdependencies between the members. In the latter strategy, there is typically a need to spend considerable time in the beginning with exercises aimed at unifying members into a work group and converging their understanding of the project into a common, albeit nuanced one, before it is possible to produce project results.

After the first two workshops of the Innovationsgym, focused respectively on individual potency and divergence, the third workshop addresses the need to unify individuals (and their individualized sense of potency) and the perspectives (exhibiting a great deal of divergence) into a work group with a social climate wherein they can begin to cooperate (with a kind of shared of group potency) and collaborate (though a kind of sharing of divergences) in the innovation process.

All workshops start with warm up exercises, some of them focused on the individual potency and some on forming individuals into a group or a team. The group-forming warm-up exercises involve everyone seeing and listening to everyone else without a designated leader. Everyone takes turns leading everyone else in making different kinds of collective rhythms and patterns of movement or sound. Finally the exercises support and encourage one another whatever he/she decides to do. There is also training in interaction and dialogue to encourage all present to both actively argue for one's own perspective and to actively listen to the perspectives of others. With this social climate of unification as a base, the different individual visions of the project begin to converge and a common vision begins to be formulated.

Generating Possibilities by Tugging *Autonomy*

When the group or organization is formed and the coarse-grained structures that define the group or organization have emerged, there is suddenly something for the individual to be autonomous from. And autonomy is necessary for individual to generate possibilities, to search for new ideas and to start developing them by experimentation. In gathering intelligence the external world was the source of new information; in generating possibilities it is the internal world, the people in the organization, who are the source. By awakening in people the possibilities afforded through *autonomy* the leader enhances the probability that new ideas with be tested, new initiatives will be attempted, new possibilities will be explored.

Giving individuals a sense of autonomy supports experimentation, and experiments cause a lot of new information, experience and knowledge to become evident and to develop and grow in the organization. To make sure that this new knowledge does not stay buried within the individual who had the experience but instead spreads across the organization, it is important to find ways to make the experience explicit and reachable for others. There are information systems for this, but just as important can be that employees have meetings where they have the opportunity to talk about work experiences and learn from each other.

Individual autonomy is about expressing one's personal involvement in work both in words and in actions and doing so as an autonomous actor, but also in the context of the organization. It is about making people understand that change within the organization that pushes the organization towards increased performance, or "fitness," is needed, and more importantly that their own personal input is expected and is in fact imperative for this to happen. Companies often use customer focus and continuous improvement programs to achieve this. Employees meet with customers and receive information that can be used to increase the organization's fitness.

Significantly, these interactions also inject energy into the system as the relationships that develop between employees and the customers drive passion and commitment into the system. This is the power of autonomy within an organizational setting. Feedback on the performance at group level and the ability to compare one's performance with others also provides a driving force. Unique knowledge about the market situation may motivate an individual to push for change, and to take risks to make change happen, but this only happens if that individual feels empowered with a high degree of autonomy. This rein is tugged by making "the work" and "the mission" important to each participant *personally*, and also make it clear that to succeed, this work must be integrated into the structures that define the organization as an entity: the visions, goals and values of the organization.

In the Innovationsgym, different methods for idea generation are used to generate many new possibilities in a short time. Post-it notes are often used as a method to give the individuals autonomy in their idea giving. In the next phase the notes are collected and used in structured activities. Such activities include norms like agreeing not to censor oneself, and not to criticize others as they give their ideas. It is also critical that suggestions that come from the autonomous individual are reinforced, and those that come from the individual trying to meet the expectations of others are not.

Structuring Activities by Tugging *Institutional Constraint*

Why do we plant seeds? The deepest answer to this question is that we plant seeds because we know that seeds contain a structure and that structure is predictable. We plant seeds because we know what is likely (with a certain probability) to happen in several weeks or months. Implicitly, we recognize that there is information embedded in each seed's structure and that information implies a predictable result. In the literal case of a seed, we know this information is embedded in genetic coding that forms biological structures under expected constraints. If we find that this predictable result might be useful, we are likely to do this same thing again and again. Thus we plant seeds and wait for useful outcomes. When we do this repeatedly in the same places and at the same times, we also structure our own activities, the "annual planting of seeds" become habit or custom, perhaps even reinforced with a festival.

When actors structure activities, they embed information in that structure just as genetic information is embedded in the chemical structure of DNA. Importantly, others can recognize this information as they see the planter spreading her seed, and then begin to stock their shelves with fertilizes and the wares necessary for the coming harvest. The patterns continue to feedback and to feed off one another, organizing the actions of many people over vast stretches of time and space. This is just as in biology when certain molecules "recognize" structures and react to (or with) them, and so on (Hoffmann, 2012). Thus, we say that stored information generates predictability. Predictability simplifies community life.

Beyond this basic level of information use, however, divergent, tacit knowledge within individuals that is informed by the information they have observed can also be externalized by transforming it into explicit and communicable knowledge. To accomplish this, a reflection process is required wherein abstract concepts are developed through interaction, and wherein different people may advocate from their individual perspective of their own internal abstractions for their particular unique understanding of the situation. However, once disconnected from the physical genesis of the information—the original observation that, yes indeed, a certain farmer is planted seeds in a particular location at a particular time—the veridicality of the information depends upon how it is reflected in the knowledge of the individual who is externalizing it.

A political process ensues. This typically ends with the selection of certain consensus points-of-view (by various groups) as dominant interpretations of the situation emerge. This process also involves making decisions about which information will be used, and in what ways it will be used. Does it matter where the farmer was? Does it matter who she is? Was she successful in the past? Or does it matter which seed was being sown? Does it even matter if there really was a farmer? Perhaps there is only a legend, but the predictability remains. Nascent structures that are based upon dominant interpretations of events are formed (See Edwards & Baker, 2013); activities converge.

Once these structures have formed around us, there is pressure on us to follow the dominant structure which becomes a constraint on future activities. But it also means that we "know" what to do in a sense, and we can start to work toward some projected purpose. By organizing activities through a common model, these structures promise to give us resources (the anticipated actions of others) and the possibility to cooperate.

This structuring part of a creative process, where one's ideas are made explicit and where certain ideas are selected for common use, is the toughest part of the creative process mentally. Individuals are forced to throw away ideas they really liked. Gathering information and generating ideas are often energizing and fun. But selecting the ideas that will go forward and then limiting the possibilities and forcing oneself to be constrained by those selected, forgoing all others is like climbing a mountain top. The footholds become more limited, the options fewer. This convergence implies *constraint.* When this phase is over, however, the path is downhill; the stage is set and the next phase of creation work can begin which is often once again very rewarding.

This part of the training in the Innovationsgym is about structuring the ideas from idea generation, determining which ones are similar and which ones are most deviant. The next phase is looking for possibilities to combine different ideas into a completely new one. Then there is the process to assess ideas and select some of them as the most promising ones.

Executing Plans and Programs by Tugging *Integration*

Executing plans and programs is the last phase of a development process. Everything is finished; the new structures are there ready to use. People have the possibility to integrate to these new structures. It can be seen as a training or exercise process where the explicit knowledge that makes up the new structures becomes tacit knowledge in the people using them. The new structures are internalized by people. People are integrated into the structures of the work place. In this way it also opens up for further development. Individuals no longer have to give resources to the structure that they have internalized. They can begin to use their slack resources to try new experiments, and the cycle can begin again. The executing function is not addressed in the Innovationsgym because the individuals who participate in the workshops take their projects back to their home organizations for implementation. The importance of this function should not be overlooked, however, because successful innovation is not realized unless the programs are executed effectively.

It can be useful to have a bigger perspective than the stand alone development process. The implementation of new structures is not the end but the start of a new development process. Each development process is like a small circle, and the most important thing is not these small circles but the series of circles, a spiral of adaptation and exploitation. The intent is to learn from each development process and be able to make them better and better. To take the opportunity to reflect and learn after each development process may be more important than to use time to plan the next.

Before we move to the conclusion, we offer one final caution: In the above discussion we focused on single rein as critical when rising to meet each leadership challenge. This is an over-simplification. Successful leadership requires that all six reins be loosened and tugged with constant vigilance. In today's organizations, circumstances change quickly, and leaders must be ready for the terrain that suddenly appears around the next bend. The ride really does have the feel—and the excitement—that one experiences vicariously when cheering for the matinee hero. This time, however, you are not at in a theater, and you are not watching a movie. It is real life.

TOWARD A COMPLEXITY-INFORMED PHILOSOPHY OF LEADERSHIP

Although it might not be immediately apparent to the reader, the perspectives we advocate in this chapter are derived from a complex systems view of human interaction dynamics (Hazy & Ashley, 2011; Hazy & Backström, 2013; Backström, 2013). Like the complex systems that have been studied in the natural sciences, we see human beings as interacting within a constrained environment, and we see human organizations as emerging from within these interactions in the context of those factors that constrain them. There is a difference however. In the case of human interaction dynamics (HID), the constraints that must be navigated are made even more complex because human agents interact in a "container" that operates on many, many dimensions, and it is made up of constraints that are both physical and social, both local and global, and both spatial and temporal.

The "reins" we have described are in effect projections of the constraints that enable and limit degrees-of-freedom along known dimensions (that are therefore recognizable), whereas other (but as yet unrecognizable) dimensions may also be relevant. But these other dimensions and the inevitable constraints they imply only come into focus over time. Dualities, we argue help us look for and potentially recognize these emerging factors and the ways they might constrain us. By finding and identifying them before they are fully formed, we can, perhaps overcome, exploit, or even change them.

The human case is further complicated because human beings often mix experiential learning which can be idiosyncratic, with social learning which might be stale and even downright false. This is especially problematic since we also have long memories that blur the distinctions between what is observed and what is learned from others. We also have the capacity to project this potentially confused mixture far into the future by communicating it with others and furthering its acceptance through political processes which often explicitly ignore the need for experiential validation of the interpretation.

Patterns in observations must therefore be found not only along each single dimension of experience, but also along its dual, that is, in how each dimension changes experientially over time. Mathematicians call these the *symplectic* dimensions. (The word "symplectic" is a cognate of the word "complex" but derived from the Greek roots rather than the Latin.) As an example, the story of a farmer who long ago planted seeds in a certain field may no longer be relevant if that field is now a desert. However, the observation that farmers plant seeds in different fields each season and at different times as latitude changes may remain relevant if the pattern continues in the present. How events were once interpreted is no longer relevant if there is evidence that events are changing and new interpretations are warranted. But the pattern from the past may remain a pattern in the present day. One must recall that the dominant interpretation of an event is a political result and its veracity is not absolute. In fact, the dominant interpretation of the event is only imperfectly related to the event itself because it occurred with limited understanding of the circumstances that surrounded it. Most of that nuanced information was lost during the politics of the convergence process.

Likewise, when wielding the reins of leadership, the pole of each duality, like Autonomy for example, is not constant but changes with regards its complement: Integration; likewise integration is not constant but changes with respect to Autonomy. The same can be said for each categorical duality. Leaders have their affect by thoughtful observation of both idealized states at the same time, of their overlaps and tensions, and of their rates of change. Leaders succeed through understanding, but also through the skillful practice in handling of the reins of leadership as others likewise search for themselves within the promise and the contradictions inherent within these dualities. Thus, the reins of leadership offer a powerful connection—perhaps our only connection—to the fates of others, as well as, our own.

ACKNOWLEDGEMENT

This research was supported in part by a grant from the Knowledge Foundation of Sweden.

NOTE

1. Correspondence concerning this article should be addressed to James K. Hazy, Department of Management, Marketing and Decision Science, Adelphi University, 1 South Avenue, Garden City, New York, 11530. Contact: hazy@adelphi.edu

REFERENCES

Backström, T. (2013). Managerial rein control and the rheo task of leadership. *Emergence; Complexity and Organization (E:CO)*, *15*(4).

Barnard, C. I. (1938). *The functions of the executive.* Cambridge, MA: Harvard University Press.

Bass, B. M. (1990). *Bass & Stogdill's Handbook of leadership* (3rd ed.). New York: The Free Press.

Bass, B. M. (1985). *Leadership and performance beyond expectations.* New York: The Free Press.

Conger, J. A. (1989). *The charismatic leader.* New York: Jossey-Bass Publications.

Edwards, M., & Baker, E. (2013). Construction in human interaction dynamics: Organizing mechanisms, strategic ambiguity, and interpretive dominance. *Emergence; Complexity and Organization (E:CO)*,*15*(4).

Goldstein, J., Hazy, J. K., & Lichtenstein, B. (2010). *Complexity and the nexus of leadership: Leveraging nonlinear science to create ecologies of innovation.* Englewood Cliffs: Palgrave Macmillan.

Goleman, D. Boyatzis, R., & McKee, A. (2002). *Primal leadership: Realizing the power of emotional intelligence.* Boston: Harvard Business School Press.

Graen, G., & Uhl-Bien, M. (1995). Relationship-based approach to leadership: Development of leader-member exchange (LMX) Theory of Leadership over 25 Years: Applying a Multi-Level Multi-Domain Perspective. *Management Department Faculty Publications.* Paper 57. http://digitalcommons.unl.edu/managementfacpub/57

Hazy, J. K. (2011). Parsing the influential increment in the language of complexity: Uncovering the Systemic Mechanisms of Leadership Influence. *International Journal of Complexity in Leadership and Management, 1*(2), 164–191.

Hazy, J. K. (2012). Leading large organizations: Adaptation by changing scale-free influence process structures in social networks. *International Journal of Complexity in Leadership and Management, 2*(1/2), 52–73.

Hazy, J. K., & Ashley, A. (2011). Unfolding the future: Bifurcation in organizing form and emergence in social systems. *Emergence: Complexity and Organization, 13*(3), 58–80.

Hazy, J. K., & Backström, T. (2013). Human interaction dynamics (HID): Foundations, definitions and directions. *Emergence: Complexity and Organization (E:CO), 15*(4).

Hazy, J. K., & Uhl-Bien, M. (2013a). Towards operationalizing complexity leadership. *Leadership.* (Available online November 25, 2013).

Hazy, J. K., & Uhl-Bien, M. (2013b). Changing the rules: The implications of complexity science for leadership research and practice. In D. David (Ed.), *The Oxford handbook of leadership and organizations.* (Available online December 17, 2013).

Hoffmann, P. M. (2012). *Life's ratchet: How molecular machines extract order from chaos.* New York: Basic Books.

Katz, D., & Kahn, R. L. (1966). *The social psychology of organizations.* New York: Wiley.

Mintzberg, H. (1981). What is planning anyway? *Strategic Management Journal, 2*(2), 319–324.

Mintzberg, H. (1990). The design school: Reconsidering the basic premises of strategic management. *Strategic Management Journal, 11*(3), 171–195.

Sashkin, M., & Sashkin, M. G. (2002). *Leadership that matters: The critical factors for making a difference in peoples' lives and organizations' success.* San Francisco: Berrett Koehler Publishers Inc.

Sun Tzu, (tr. Cleary, T). (2005). *The art of war.* Boston: Shambhala Publications, Inc.

Surie, G., & Hazy, J. K. (2006). Generative leadership: Innovation in complex environments. *Emergence: Complexity & Organization, 8*(4), 13–26.

Weber, M. (1946). *From Max Weber: Essay in sociology* (H. H. Gerth & C.W. Mills, eds). New York: Oxford University press.

Weick, K. E. (1995). *Sensemaking in Organizations.* Thousand Oaks, CA: Sage Publications.

PART IV

NEW LANDS

CHAPTER 8

FOUNDATIONS FOR CROSS-CULTURAL SUCCESS

Deborah E. Gibbons
Naval Postgraduate School

ABSTRACT

Cultural predispositions impact our interactions, partnerships, working styles, and modes of organizing. Cultural preferences affect the way we see the world and the way the world sees us. For people who operate in a variety of cultures, success depends on the ability to sustain core values and goals while adapting to local beliefs and practices. In some situations, managers need in-depth understanding of one local culture. In other situations, managers need to work with a variety of international stakeholders, each of whom carries his or her own cultural imprint. Long-term effectiveness across ethnic, religious, or national boundaries requires that leaders and managers develop cultural competencies that will enable them to adapt appropriately. These competencies include the ability to analyze cultural tendencies, resolve human rights issues, and communicate with divergent audiences. Necessary skills include distinguishing between innate-human and cultural factors, interpreting key dimensions of culture, and knowing which kinds of cultural attributes require special attention. Cultural competencies empower organization leaders to build positive partnerships and effective organizational processes in cultures that are not their own. This chapter will help you develop your cultural com-

Millennial Spring: Designing the Future of Organizations, pages 147–163
Copyright © 2014 by Information Age Publishing
All rights of reproduction in any form reserved.
147

petencies. Before you reach the final page, you will know more about your cultural predispositions, about managing successfully across cultures, and about helping your organization make a positive impact on its international employees, partners, and communities.

INTRODUCTION

If you travel regularly, you know that cultural differences have significant effects on everything from queuing (or not) at the customs station to coordinating with local partners. You may have experienced an awkward moment when you went for a two-cheek kiss and found that locals only kiss one, or perhaps you forgot to bring tissue to the bathroom in a country where none is provided. Courtesies are best learned before you go, including "hello, thank you, excuse me," and "where is . . . ," but you can pick them up quickly onsite if you must. What you cannot pick up quickly is cultural competency, the ability to navigate successfully among people and organizations that may not share your cultural background. Unlike "please" and "thank you," which are very specific, cultural competencies are general skills that you can apply to each situation as you interact, communicate, and build relationships.

Travel websites and consulting firms sometimes offer advice about negotiation, politeness, and other behavioral guidelines. These worthwhile tips may contribute to reaching your goals when working outside your home culture, but they do not prepare you for the deeper aspects of inter-cultural interaction. Long-term success across national, religious, or ethnic borders requires that leaders and managers understand key types and effects of cultural values. You need to know how culture impacts behaviors, and you need to know how to adjust your words and actions so they will be effective. Cultural competency depends on the ability to analyze cultural tendencies and discern when cultural differences require special consideration.

Foundations for cultural competency include distinguishing between innate-human and cultural factors, understanding cultural dimensions and their effects on people and groups, and recognizing human rights issues that modern organizations must address as they expand across borders. Cultural competencies enable organization leaders to adjust their interactions and strategies to build positive partnerships and effective organizational processes. In this chapter, we will consider a few exemplary circumstances and build foundations for cultural competency. Through the process, you will have the opportunity to learn about your own cultural biases, about managing successfully across cultures, and about helping your organization make a positive impact on its international employees, partners, and communities.

Sources, Types, and Effects of Cultural Values

We can define culture as a man-made set of shared behaviors, norms, values, and assumptions that develop within a social system and are taught, both actively and passively, to new members of the system (for related definitions, see Hofstede, 2001; Schein, 2004). Cultural values are standards and preferences that groups of people come to share. They often arise from religious beliefs that indicate right and wrong, from environmental pressures that foster new behaviors, or from social interventions that indoctrinate an upcoming generation. The values are supported and transmitted through social relations and modeling, education systems, and ongoing communication. Sometimes the source of a shared value is forgotten while the value remains. The values influence norms—social rules about how things ought to be done—and together, the values and norms constrain behaviors.

Cultural values tend to be fairly stable, but they can change from one generation to the next as new people and ideas enter the social system, or as people and ideas associated with the former culture are removed from the mainstream. For example, after the Soviet Union annexed smaller nations in the twentieth century, they systematically indoctrinated people with state-approved beliefs and values. Even after many years of independence, we find similarities in cultural values among formerly Communist countries. Perhaps not surprisingly, people in these countries still tend toward secular values and place stronger emphasis on survival than on self-expression (Inglehart & Welzel, 2005, 2010).

Innate Needs, Culture, and Human Rights

While all people share basic needs, we do not all share the same values. You are probably familiar with Maslow's (1943) theory that people are motivated by crucial categories of human needs. These categories initially included physiological needs, safety, love and belonging, esteem, and self-actualization. While these categories of needs were clearly relevant to motivation and happiness, they were not all-inclusive. Recognizing significant gaps in his original model, Maslow (published posthumously in 1971) completed his work by adding "self-transcendence" and "peak experiences" as the ultimate human needs. In this revision of his theory, Maslow removed the faulty assumption that people necessarily seek to meet lower-level needs before pursuing higher-level needs, and he acknowledged the potential for spiritual desires to overshadow all other needs.

Building on Maslow's research, Koltko-Rivera's more recent work clarifies the motivational relationships among needs: "The crucial issue is the dominant motivation at work in the individual's life. All individuals experience hunger; however, hunger is the defining experience only for individuals centered on the physiological or survival level of Maslow's hierarchy of needs" (Koltko-Rivera, 2006, p. 307). With these adjustments in place, Maslow's and Koltko-Rivera's works provide foundations for understanding the panorama of human needs that shape individuals' satisfaction and well-being. Similarly, Lawrence and Nohria (2002) proposed that humanity is driven by needs to acquire things and status, to defend ourselves, to learn about the world around us, and to bond with others. Basic needs affect people from all cultures, but different social groups place greater emphasis on some needs than on others.

Cultural values often relate to basic needs, governing how certain needs should be met or perhaps how they should not be met. Many cultures share the belief that people have a right to physical safety, and many cultures disapprove of theft, rape, slavery, and other behaviors that meet one person's needs at the expense of another person. Other cultures define which groups are acceptable targets for such behaviors and which groups should be protected. Such cultural values explain how groups of people can sometimes justify actions that devastate members of different social groups, as we have seen in Sudan, Rwanda, Pakistan, and Nigeria in recent years. It is important for people who work internationally to be aware of cultural values that pertain to human rights, and to take steps to uphold international standards within our own spheres of influence. In this regard, I have found it very useful to refer to the Universal Declaration of Human Rights (United Nations, 1948) as the cross-cultural standard when I am working with people whose cultures do not consistently uphold human rights. Because this document was created by the international community, it provides good footing for organizational policies that protect human rights.

The United Nations adopted the Universal Declaration of Human Rights via General Assembly resolution 217 A (III) of 10 December, 1948. The preamble states that "recognition of the inherent dignity and of the equal and inalienable rights of all members of the human family is the foundation of freedom, justice and peace in the world..." The Universal Declaration of Human Rights includes 30 articles, the first two of which establish the equality of all people:

> *Article 1.* All human beings are born free and equal in dignity and rights. They are endowed with reason and conscience and should act towards one another in a spirit of brotherhood.

> *Article 2.* Everyone is entitled to all the rights and freedoms set forth in this Declaration, without distinction of any kind, such as race, colour, sex, language, religion, political or other opinion, national or social origin, property, birth or other status. Furthermore, no distinction shall be made on the basis of the political, jurisdictional or international status of the country or terri-

tory to which a person belongs, whether it be independent, trust, non-self-governing or under any other limitation of sovereignty.

Subsequent articles address freedoms of speech, assembly, religion, movement, marriage, work, health, education, physical safety, equality before the law, and others. The Proclamation of Teheran, adopted by the International Conference on Human Rights (1968) in Iran reiterated that "the Universal Declaration of Human Rights states a common understanding of the peoples of the world concerning the inalienable and inviolable rights of all members of the human family and constitutes an obligation for the members of the international community." Many cultures do not understand human rights, and these international documents give us a good foundation for educating and inspiring people from all cultures to defend everyone's rights.

Cultural Dimensions and Their Effects on People and Groups

Whether you are working in Asia, Latin America, Europe, or elsewhere, you can benefit from understanding several dimensions of culture. These cultural dimensions give us information about the general tendencies of a population, but they do not tell us exactly how a particular person feels. The best known and validated measures have been tested through extensive surveying of people in many countries, and the results are aggregated by country or by society. Researchers from a variety of nations have contributed to this work, providing helpful insights about cultural tendencies around the globe.

Hofstede (2001) used a large sample of international IBM workers to identify five key dimensions of culture, including individualism versus collectivism, acceptance of power differential in society, uncertainty avoidance, distinction between masculine and feminine roles, and long- versus short-term orientation. Each of these cultural dimensions affects the way people make choices and interact with each other. For example, during a training session for UN peacekeepers in Jordan, I asked how they might help reduce friction between locals and refugees. A woman replied that if she had a dispute with one of her neighbors, her sister would go to that person and work it out. This is a collectivist response. People from individualistic cultures would be more likely to outline procedures for mediating disagreements without relying as heavily on their social networks. In general, collectivist societies see themselves in terms of family and friends, and they often prefer to do business with friends of family members or families of friends. Working within their networks enables them to trust the other people, even as it requires them to fulfill their obligations toward other network members. Individualists may use social networks to accomplish their goals, but they are less likely to define themselves in terms of their relationships and memberships, and they are more likely to rely on explicit contracts to govern business transactions.

Acceptance of power differential, which Hofstede named "power distance" can impact effectiveness in organizations. High power distance means that members of the culture accept large discrepancies in power and wealth, such that both the powerful and the powerless are tolerant of this inequality. Low power distance means that members of the culture value equality, and they do not believe that some people should have dramatically more power or wealth than others. People from high power-distance cultures expect supervisors to be bossy and workers to be submissive. People from low power-distance cultures do not. How do you suppose that the United States and Canada compare to other parts of the world in this regard? Thinking about the way Israeli people debate politics and the way their Arabic neighbors debate politics; which culture do you imagine is very low on power distance and which is very high?

If you guessed that the United States and Canada have a fairly low tolerance for power discrepancies, you are correct. Israel, however, has even less tolerance for power discrepancies, ranking as one of the lowest power-distance nations in the world. The Arabic cultures, in contrast, have a long history of power differential, which has largely been accepted as normal by members of their societies. In the Hofstede studies, the Arabic countries ranked seventh in the world for power distance. Although we now see rising demands for civil rights among some members of these societies, the structures of emerging governments are likely to retain large power distances. Nevertheless, increasing access to international values and practices could open these cultures to more egalitarian ideas and systems.

Two more aspects of culture that affect people's decision-making and risk-taking are uncertainty avoidance and long- versus short-term orientation. Cultures that foster discomfort regarding ambiguity are high on uncertainty avoidance. In these countries, people tend to lack confidence in their ability to influence politics; they tend to have more corruption and more laws; and they tend to be more suspicious of their governments (Hofstede, Hofstede, & Minkov, 2010). They generally rely on social norms and formal procedures to structure behaviors. Interestingly, in a study of help for strangers, people from the uncertainty-avoiding cultures were more likely to help a blind person cross the street than people from cultures that are more tolerant of uncertainty (Levine, Norenzayan, & Philbrick, 2001). Long-term versus short-term orientation has often been used to explain some of the differences between Asian and American business practices. Asian cultures tend to take a long-term approach, believing in self-discipline, social cohesion, concrete thinking, and persistence. They tend to hold relativistic values and to focus on life-long planning. Many European cultures take a more short-term approach, believing in personal achievement, meritocracy, universal truths, and rationality. They tend to hold fixed values and to focus on immediate payoffs. These differences can produce very different strategies, as the long-term orientation leads people to seek stability while the short-term orientation leads them to seek speedy profits.

Masculinity versus femininity measures the extent to which a culture designates separate roles for men and for women. A high masculinity score indicates a culture in which male roles are distinct from female roles. A high femininity score indicates that women and men are treated as equals with little differentiation in their roles within society. Sometimes this distinction can create dramatic differences within organizations. For example, I work for the United States Navy, where women often face challenges entering the traditionally male organization. The United States is somewhat high in masculinity. While visiting peacekeepers recently in Chile, which is high in femininity, I asked a female Army major what challenges and opportunities she had encountered in her military career. She said that everything had been fine since the military academy opened to women. Surprised by her extremely positive story, I asked what obstacles she had to overcome. With a trace of irritation, she insisted that there had been no problems. Knowing that her father was a high-ranking officer, I wondered if family connections had streamlined her career. A couple of days later, I asked another female officer what challenges women face in the Chilean army. Clearly offended, she asked the interpreter, "Does she think that we are backwards?" Flummoxed, I asked a Central American instructor, who works extensively with peacekeepers throughout the Americas, if women in the Chilean army do not experience the kinds of difficulties that United States women have had. He replied that there are problems with gender inequalities and sexual harassment throughout Latin America ... with the possible exception of Chile, where women usually receive respect. Another colleague later explained that men in the Chilean army rarely behave inappropriately toward female colleagues, but if one did, he would be dismissed immediately. Cultural differences between the United States and Chile appear to have formed discrepant foundations for integration of women into the two countries' military organizations. My cultural competency in this situation was low because I assumed similarity between military cultures. In reality, an organization's culture reflects the country in which it was founded, and in this case, Chilean cultural attributes blended with the military structure to create one of the most positive environments for military women in the Americas.

Another notable stream of research on national cultures has been the Global Leadership and Organizational Behavior Effectiveness (GLOBE) study. The research measured values and corresponding practices in 62 societies within 58 countries, identifying cultural dimensions similar to Hofstede's plus three additional dimensions that may be useful for people working across cultures (House, Hanges, Javidan, & Dorfman, 2004). The three additional dimensions—assertiveness, humane orientation, and performance orientation—vary by culture. Assertiveness is practiced and valued to different extents, with some cultures indicating a preference for more assertiveness and others for less than their current practice. Humane orientation indicates the extent to which people are rewarded for showing kindness and generosity. Most participants in the GLOBE study reported stronger humane values than they experience in practice. Performance orientation indicates the extent to which people are rewarded for improvement and excellence in their performance. On average, respondents to the GLOBE surveys desired stronger emphasis on performance-based rewards than they experience in practice. In addition, those who were more performance oriented were also more inclined to look for charismatic leaders who would bring change in their organizations. Understanding the level of a culture's support for these values can help us predict the natural inclination of co-workers and business partners so we will not be blindsided by unmet expectations.

The World Values Survey is another ongoing, broad-scale sequence of surveys that measures social, economic, demographic, and attitudinal variables in many different countries. Minkov replicated Hofstede's findings with these data and added a new dimension that he named "indulgence versus restraint" (Hofstede, Hofstede, & Minkov, 2010). This dimension combines two items: perceived control over one's life and importance of leisure time. According to Minkov, a higher score indicates greater indulgence of personal desires within society, and a lower score indicates more societal restraint. To see the distribution of indulgence versus restraint across 41 countries, I analyzed World Values Survey data that were collected in the year 2000. In the same analysis, I measured long-term orientation. Figure 8.1 shows the results, with countries placed in the graph according to their scores on indulgence versus restraint (y-axis) and long-term orientation (x-axis). In this graph, we find several Latin American countries at the top left, scoring high on indulgence (perceived control and importance of leisure) and low on long-term orientation (indicating that they hold fixed values and live for now). At the opposite corner of the graph, we see a cluster of countries that were heavily influenced by the Soviet Union, and as of A.D. 2000, they continued to feel little control over their lives, place little emphasis on leisure, and focus on long-term planning, social cohesion, concrete thinking, and persistence. If you look carefully, you will be able to find Asian countries that are likewise high on long-term orientation, but not as personally restrained as the former-Soviet countries. Can you

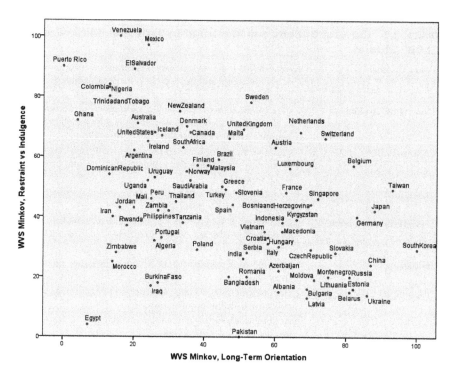

Figure 8.1 Countries sorted by levels of indulgence and long-term orientation. *Source:* World Values Survey data, A.D. 2000; larger values indicate higher levels of indulgence/long-term orientation using Minkov's measures of indulgence and long-term orientation

find places that you have visited? Would you have predicted these scores for people from those countries?

The Hofstede, Minkov, and GLOBE dimensions of culture are summarized in Table 8.1. As you read through the table, ask yourself where you fit in each dimension, and if others from your home culture seem to share your tendencies. Have you worked with someone from a different culture who seemed to expect (and deliver) more or less equality among workers than you are accustomed to having? If you worked with people from Mexico, where power distance is high, you would probably find more deference to supervisors and community leaders than among people who grew up in the United States. If people from Sweden, where power distance is very low, joined your working group, you would probably find very little status differentiation. These distinctions affect the way people work together, and they can create misunderstandings when we fail to realize that our colleagues see the social world from very different perspectives.

TABLE 8.1 Cultural Dimensions Identified by Hofstede, Minkov, and GLOBE Studies

Power Distance*	Extent to which the members of organizations, institutions or society accept and expect that power is distributed unequally
Uncertainty Avoidance*	Extent to which the members of organizations, institutions or society feel uncomfortable with ambiguity, which they minimize using social norms, laws, or procedures
Collectivism*	Extent to which people's identities and loyalties are invested in cohesive social units that regulate behaviors and provide social support
Masculinity*	Extent to which men and women are assigned to distinct gender-based roles by society
Long-Term Orientation*	Extent to which society values pragmatic virtues, such as saving and persistence, which are oriented towards future rewards
Indulgence**	Extent to which people are free to choose how they spend their time
Institutional Collectivism***	Extent to which society encourages collective action and collective distribution of resources
Assertiveness***	Extent to which society encourages individuals to behave assertively or to confront others
Humane Orientation***	Extent to which society encourages humane behaviors such as altruism and generosity
Performance Orientation***	Extent to which society encourages rewards based on personal performance

* Hofstede dimension with GLOBE counterpart, ** Minkov dimension, *** GLOBE dimension

Inglehart and Welzel (2005, 2010) used World Values Survey data to distill entire cultures down to two variables. One measure places people along a continuum anchored at one end by concern about survival and at the opposite end by desire for self-expression. The other measures a continuum between traditional values and secular values. The Hofstede, Minkov, and GLOBE dimensions differ in character from these Inglehart and Welzel dimensions. Concepts like power distance or individualism are singular and basic. Concepts like focus on survival or traditional values reflect underlying, more fundamental processes such as a recent history of hardship (survival focus) or commitment to monotheistic faith (traditional values). Not surprisingly, people from countries that have experienced a lot of poverty, violence, or repression value survival more than self-expression, while wealthier and freer countries sit near the self-expression end of this measure. Countries where people practice Christian or Muslim faiths tend to believe in the right to life and traditional marriage, while countries where people practice other religions or no religion tend to support abortion, euthanasia, non-traditional marriage, and divorce. Can you guess how the United States and Western

Europe differ on these measures? Both value self-expression, but the United States holds more closely to traditional values while much of Western Europe (except Ireland) holds more secular values.

Cultural Attributes That Require Special Attention

Some aspects of culture require particular attention because they are likely to affect our interactions. If we behave more humanely or reward performance a bit more than is customary in the local culture, we are not likely to suffer backlash because most cultures believe that these behaviors should happen more than they do. In contrast, if we ignore collectivism, uncertainty avoidance, power distance, gender roles, or assertiveness, we could make serious mistakes. The GLOBE study found that certain leadership attributes were valued in all cultures, while others were seen as positive in some cultures and negative in others. A charismatic, value-based style of leadership finds approval worldwide, as does team-based leadership. Several leadership attributes, such as logic, risk-taking, and ambition, are valued in some cultures and disapproved in others (House, Hanges, Javidan, & Dorfman, 2004). We should not rely on our own beliefs in these areas. Rather, we can emphasize universally-positive behaviors while we watch for cues about the culturally contingent behaviors.

When we begin a relationship with someone from another culture, we cannot assume that they will welcome us as readily as we welcome them. Ethnocentrism, or preference for people from our own social/cultural group, can have very strong effects on attitudes and behaviors. We have seen extreme cases in tribal warfare, genocide, dhimmitude, and caste systems, but the more subtle effects on our daily interactions should not be overlooked. We must expect that many people, perhaps most people, beyond our Western culture will favor insiders with whom they identify socially and culturally, and they may feel little or no responsibility toward outsiders. This is more likely to occur in collectivist countries. Consider a Hispanic-American woman's story about in-group and out-group prices in Mexico:

> My husband, an Anglo-American, and I, a Hispanic-American, decided to go to the market to buy a few souvenirs. We got separated while browsing. He found a tapestry that he just had to have. When he asked the young lady how much it was, she told him 65 dollars. He thought, okay, let me think about this. We finally catch up with one another. He tells me what he wanted and tells me the price. I am like "no way." So I leave him once again and proceed to ask the same young lady about the tapestry, only I ask in Spanish. She proceeds to tell me it is 30 dollars. About this time, my husband walks up and the young lady realizes we are together. The redness in her face indicated she knew what had just occurred. I did not proceed to ask or embarrass her about

the price quote difference. In her country, the act of bartering is common place. But why was my price quote considerably lower than his? I believe that it was strictly because of our cultural likeness. My appearance gave away the fact that I have Mexican heritage. The fact that I speak Spanish passably was sure to have helped.

Collectivism is a particularly potent aspect of culture, and people from collectivist cultures see the world very differently from the view that is common in more individualistic cultures. Combined with high power distance and typical non-assertiveness in Asia, this can nullify the value of typical American feedback and reward practices. Consider the following story, told by a physician from the United States about his experience working in Japan:

> My first surprise was the utter horror and embarrassment bestowed on the recipient of the Employee of the Month Award. This Western artifact of individual accomplishment was the antithesis of the Okinawan focus on community and teamwork. In fact there is an Okinawan word for this focus called "moai" which is a type of support network composed of friends and family who provide financial, emotional and social help throughout life. Many researchers attribute this as a key aspect of Okinawan culture that contributes to their longevity (Okinawa is known to have one of the highest percentages of centenarians in the world). We found that our monthly birthday parties with cake and tea were much more well received and in keeping with the team approach. Perhaps this strong focus on teamwork overshadowed and even masked internal conflicts that only became apparent when we attempted to change work groups or schedules. The language barrier likely decreased our ability to pick up on internal squabbles and rivalries as well ... After living in Japan for almost three years, I found myself becoming acculturated. I found myself enjoying the reverence for politeness and respect. Very rarely did I experience indifferent or rude behavior. Quite often, I was struck by the way people went out of their way to provide assistance. Most of the time, this made coming back to the United States an uncomfortable culture shock where rudeness and indifference can be considered in no short supply. I also had to keep in mind that this cultural norm could possibly prevent workers from speaking up when they were dissatisfied, for fear of the appearance of rudeness or disrespect. I tried to make a point of asking for a lot of feedback in an attempt to improve work processes. It seemed to help. I found that over time, the workers were more comfortable with providing suggestions for improving their work processes.

In keeping with the collectivist aspect of Asian cultures, Asian managers are more likely than Americans to attribute failures to entire organizations rather than to individuals (Friedman, Liu, Chen, & Chi, 2007). Similarly, people from Asian cultures are more likely than people from more individualistic cultures to make group-based stereotypes, whether good or bad (Spencer-Rodgers, Williams, Hamilton, Peng, & Wang, 2007). Nevertheless, people from collectivist cultures share the individualist's desire for a reward system at work that pays effective workers more for doing a good job (Fischer & Smith, 2003).

Gender roles, especially when they are intertwined with religious practice, can create awkward or impossible situations: For example, a team of American men extends handshakes to head-scarfed, uniformed, military women in Jordan. One of the Jordanian women retracts her right hand, fingers curled against her chest. An American woman does the same with a bearded businessman in Israel. The Israeli man clasps his hands behind him. In both situations, the Americans misread modern clothing as an indicator of gender equivalence. Another example: An international team builds a new elementary school in a Muslim-dominant region, but after a few days of classes, many of the little girls are kept at home because the girls' and boys' bathrooms are located side-by-side. Pragmatic plumbing cannot stand against parents' concerns for their daughters' virtue, and new bathrooms must be constructed. Local people would not have made these mistakes. International organizations have an ethical obligation to provide equal treatment to men and women, boys and girls, but we often need to adjust our behaviors and services to meet the cultural realities within each country. Consultation with a local expert regarding gender issues is often advisable.

Despite our differences, expectations of good leaders are similar in many ways across cultures. For example, the GLOBE study demonstrated cross-cultural support for transformational leadership, including an overarching vision, encouragement, trustworthiness, being positive and proactive, planning for implementation of the vision, and motivation of employees toward the vision (Carl & Javidan, 2001). Nevertheless, the way the vision is communicated, and the way people are encouraged to engage that vision, must be tailored to the particular cultural setting.

Communication across Cultures

A British colleague suggests in perfect English that we begin our syndicate work at half eight in the morning. The Americans huddle to decide how a syndicate might relate to their team, and where or when is half eight.

The Japanese prospective-partner greets his Canadian guests at the airport, and despite their time limitations, insists on taking them sightseeing. The tours, teas, and dinners are wonderful, but the pleasantries take so long that the Canadians have to extend their stay to complete the negotiations and write a contract.

Culture shapes the way people communicate. Even between British and United States speakers, we find amusing differences in word choice, posture, and use of formalities. As we interact with cultures that are very different from our own, language barriers are amplified by fundamental differences in the way we communicate. Some cultures favor direct communication, with precise wording and all of the details that are necessary for understanding. Other cultures prefer nuance, reading between the lines, use of body language, and subtle linkages to shared cultural knowledge. Hall (1976) identified the former as "low-context" cultures and the latter as "high-context" cultures. In high-context cultures, as in Asia and Africa, people may avoid speaking directly. This places the burden of understanding on the hearer's ability to pick up contextual clues and interpret cultural references. Successful communication in a high-context culture depends on understanding of the norms, values, and shared experiences within that (generally collectivist) culture. Status is important for credibility, but the highest-ranking person may not take the stage and make demands. Rather, people will be attuned to his status and willing to follow his suggestions. Relation-building, courtesy, and trust are important aspects of communication, and people try very hard to avoid rudeness. For example, a Japanese person may find several ways to avoid saying "No" because she does not want to embarrass anyone with the refusal. If she makes an excuse to leave the room, changes the subject, or gives a vague yes-I-understand-you response, she probably means "No," but if her English-speaking listener is not attuned to the cultural embargo against a rude, direct refusal, he may not get the message. In low-context cultures, as in the English-speaking and Scandinavian countries, people tend to say what they mean as clearly and accurately as possible. The low-context person also tends to interpret others' communications in the same way. This can lead to misunderstandings when a low-context hearer interprets the words of a high-context speaker.

When you want to communicate with someone from an unfamiliar culture, how should you begin? First, do not assume that they see the world as you do, or that they communicate like you do. Second, try to listen more than you speak. Watch for clues and hints, especially if they are from a high-context culture. Listen actively, and do not rush to conclusions. As you begin to comprehend others' perspectives, try to frame your thoughts in terms of their interests and experiences. Adler (2002) suggests that cross-cultural communication will go better when we assume differences until similarity is proven, emphasize description rather than evaluating what we hear, try to understand the other person, treat our interpretations as working hypotheses, and seek advice from local people who can discern if our understanding is correct. To take this approach, you must set aside your opinions and listen without immediately trying to interpret what you hear. You might ask non-judgmental questions that invite further information-sharing. "Can you explain more about that? Could you give me an example?" Before you try to frame a response, summarize the message that you think you have heard, and check to see if the key points are right. "Do I understand correctly that...?" Most people will appreciate this effort, and they are likely to respond by confirming the parts you have right and filling in details that you have missed. After you ascertain your colleagues' thoughts, you will be better able to align your ideas with theirs and also help them see from your perspective.

Increasing Your Cultural Competency

There will always be surprises when we enter an unfamiliar culture, but practice strengthens our ability to read and respond to cultural signals. If you intend to work extensively within one culture, you might want to invest in social and cultural education. You will be better able to develop cross-cultural alliances if you understand the people's background, values, and practices while focusing on building trust, conforming to local practices, and sharing leadership (Graen, 2008). Even without formal education, you can shorten the learning process by asking others about their experiences and by discussing your observations and goals with local friends or colleagues.

Whether within a single culture or across many, you can improve your cultural competency every time you interact with people whose backgrounds differ from yours. Observe and analyze body language in group settings. How do people interact with each other? What does this tell you about gender roles, assertiveness, and power distance? As you examine others' social patterns, you may need to apply your new knowledge to find an acceptable compromise between your preferences and the prevailing local norms. For example, when I taught a course at Cairo University in Egypt, I felt dismayed by the extreme power distances between professors and the supporting staff. The director of the program where I was teaching had a wall switch behind his desk that activated a light in the hallway, and when he flipped the switch, his assistant came running. Despite my discomfort with such hierarchy, I had to adapt my behavior to this environment. Basic courtesies, which seemed to surprise some people, were appreciated across all social levels, but the staff members remained deferential throughout my six-week stay. I found that our relationships worked best when I exceeded status barriers minimally and gradually. Given more time, I could have slowly reduced the power differential within my sphere of influence, but I would still need to respect social structures that help each person to fit—and feel comfortable—within their society.

Where possible, messages should be framed to align with local norms and values. Try active listening; see if you can assess your colleagues' cultural tendencies and relate to their viewpoint. Do the people seem to take a collectivist perspective? If so, do you recognize signs of high-context communication? In such a culture, I generally adopt a speech pattern that emphasizes teams, relationships, and shared commitments rather than individual goals or accomplishments. My level of acceptance by new acquaintances increased dramatically when I began consciously taking this approach to communication. If my acquaintances talk about their families, then I do, too. If they are very formal, so am I. If they seem to focus on one or two aspects of the project that we are undertaking together, I begin my conversation from that point, and build bridges from there to the messages that I need to convey. It is useful to put yourself in the place of your international audience (even if the audience is a single person). Look for the discrepancies between their environment and yours. Continually analyze communication, striving to understand how they present ideas and which of their cultural values are the most essential. If you can find common ground, emphasize that. It is much easier to effectively present ideas to someone with whom you already have a connection; it is therefore more effective to develop a foundational relationship before trying to convince anybody of anything. After you establish a positive relationship, frame your ideas using examples, ideals, and modes of communication that reflect what you have heard from your audience. People tend to employ rhetoric that they find effective themselves; use this to your advantage. With these analysis and response skills, you can streamline your travels, create more meaningful and effective relationships, and help your organization to make a positive impact on international employees, partners, and communities.

REFERENCES

Adler, N. (2002). *International dimensions of organizational behavior* (4th ed.). Cincinnati, OH: South-Western Publishing.

Carl, E., & Javidan, M. (2001, August). Universality of charismatic leadership: A multi-nation study. *Academy of Management Proceedings*, B1–B6.

Fischer, R., & Smith, P. B. (2003). Reward allocation and culture: A meta-analysis. *Journal of Cross-Cultural Psychology, 34*(3), 251–268.

Friedman, R., Liu, W., Chen, C. C., & Chi, S. S. (2007). Causal attribution for interfirm contract violation: A comparative study of Chinese and American commercial arbitrators. *Journal of Applied Psychology, 92*(3), 856–864.

Graen, G. B. (2008). Linking Chinese leadership theory and practice to the world: leadership secrets of the Middle Kingdom. In C. C. Chen & Y. Lee (Eds.), *Leadership and management in China: Philosophies, theories, and practices* (pp. 272–297). London: Cambridge Press.

Hall, E. T. (1976). *Beyond culture.* Garden City, New York: Anchor Press.

Hofstede, G. (2001). *Culture's consequences: Comparing values, behaviors, institutions, and organizations across nations.* Thousand Oaks, CA: Sage Publications.

Hofstede, G., Hofstede, G. J., & Minkov, M. (2010). *Cultures and organizations: Software of the mind* (3rd ed.). New York: McGraw-Hill Professional.

House, R. J., Hanges, P. J., Javidan, M., Dorfman, P. W., & Gupta, V. (2004). *Leadership, culture, and organizations: The GLOBE study of 62 societies.* Thousand Oaks, CA: Sage Publications.

Inglehart, R. F., & Welzel, C. (2005). *Modernization, cultural change and democracy.* New York: Cambridge University Press.

Inglehart, R. F., & Welzel, C. (2010, June). Changing mass priorities: The link between modernization and democracy. *Perspectives on Politics, 8*(2).

Koltko-Rivera, M. E. (2006). Rediscovering the later version of Maslow's hierarchy of needs: Self-transcendence and opportunities for theory, research, and unification. *General Psychology, 10*(4), 302–317.

Lawrence, P. R., & Nohria, N. (2002). *Driven: How human nature shapes our choices.* San Francisco: Jossey-Bass.

Levine, R. V., Norenzayan, A., & Philbrick, K. (2001). Cross-cultural differences in helping strangers. *Journal of Cross-Cultural Psychology, 32*(5), 543–60.

Maslow, A. H. (1943). A theory of human motivation. *Psychological Review, 50*(4), 370–396.

Maslow, A. H. (1971). *The farther reaches of human nature.* New York: Viking Press.

Schein, E. H. (2004). *Organizational culture and leadership* (3rd ed.). San Francisco: Jossey-Bass.

Spencer-Rodgers, J., Williams, M. J., Hamilton, D. L., Peng, K., & Wang, L. (2007). Culture and group perception: Dispositional and stereotypic inferences about novel and national groups. *Journal of Personality and Social Psychology, 93*(4), 525–543.

CHAPTER 9

INTRODUCING DESIGN-DRIVEN INNOVATION

A Challenging Ride Towards New Possibilities[1]

Marcus Jahnke
Ulla Johansson Sköldberg
University of Gothenburg, Sweden

ABSTRACT

In recent years interest has grown in how design can contribute to innovation in business and society, such as through the management concept of design thinking. However, up-close studies on the contribution of design practice to innovation are scarce, and the artistic dimension of design practice, and its relevance for innovation, is often neglected. This chapter draws on an experimental study in which designers involved multi-disciplinary groups of non-designers in companies in experiencing design practice hands-on to strengthen innovation processes. The chapter gives a number of examples of what happened in this work, and argues that design-driven innovation contributes to innovation by challenging and expanding how companies understand their products. However, to introduce design-driven innovation, and in this way draw actively on interpretative and hands-on artistic skills, rather than analyt-

Millennial Spring: Designing the Future of Organizations, pages 165–190
Copyright © 2014 by Information Age Publishing
All rights of reproduction in any form reserved.

ic approaches, also means a number of challenges for managers. The chapter presents six such challenges and the management skills necessary to engage in design-driven innovation.

A CHALLENGING RIDE TOWARDS NEW POSSIBILITIES

Most people view designing as about colour and shape. But few people are aware that there is much more to design. Design researchers talk about design as a problem-solving methodology—or a way of tackling what are known as wicked problems (Rittel & Webber, 1973),[2] problems that can't be dealt with purely using analytic methods. However, when describing the role of the designer and the way designers work in the context of innovation, this problem-solving approach does not go far enough. If designers and engineers merely had different problem-solving methodologies, it would presumably be much easier for them to work together and to exchange experiences than appears to be the case.

In this chapter we assert that design must be understood as a special practice characterised by combining an interpretive and analytical approach in a creative process that may lead to innovative results. We give examples of what this may mean to companies by drawing from a recent research study. We also discuss economic and managerial implications of introducing design-driven innovation.

THEORETICAL PERSPECTIVES ON DESIGN PRACTICE

The ability of a designer to achieve innovation is attracting increasing attention. Descriptions of the work of the successful American innovation and design firm IDEO have been popularised in films and books (e.g., Kelley, 2001; Brown 2008). Much has also been written about the concept of Design Thinking in the business press (e.g., Business Week, Fast Company, The New York Times). In this context, innovation challenges are seen as complex or wicked problems that the designer's process and way of thinking are particularly well suited to solving. The view that the design process is about problem-solving can be traced back to Nobel Laureate Herbert Simon's definition from the 1960s: "Everyone designs who devises courses of actions aimed at changing existing situations into preferred ones" (Simon, 1969/1996, p. 135), a definition which is often used to describe innovation and problem-solving.

Understanding Design's Contribution to Innovation Beyond Problem-Solving

It is true that a successful innovation can often be described as such afterwards, or once a problem has been solved or when success has been achieved in the market place. However, we believe that this is a view that obscures another important way of understanding design's contribution to innovation; the designer is often working to a different, more aesthetic, rationality in which creating and materializing meaning are central to the creative process. If we take this as our starting point, we can begin the innovation journey without having a problem at all.

While the central path of innovation research developed out of Simon's theories of decision-making, research which has sought to understand the praxis of the designer has taken a different route and found knowledge elsewhere. For those interested in design as an aesthetic practice, American philosopher John Dewey's thoughts are interesting (1934). He argued that aesthetic practices, taking art as his prime example, are about translating experiences and meanings from one situation to another via some form of medium. The medium may, for example, be the painter's painting or the furniture designer's chair. While Dewey's view of art might feel old-fashioned, his emphasis on experiences and meanings is important. One could say that his philosophy ignores problems as a starting point altogether. Instead an aesthetic process can be triggered by almost anything, for example an observation of something interesting, a phenomenon that attracts attention. It might be the way light is reflected in wet asphalt as the inspiration for something that later becomes a new type of surface for a car, or an interest in how social values, such as greater gender equality, might be important to consider when developing new types of hand-held tools. Whatever the focus of interest may be, the thing that may later become a new idea grows out of a vague or ambiguous situation rather than out of an actual and well-defined problem. In order to engage in such situations, in homing in on what is important about the situation, it is necessary to engage in meaning making—active interpretation to attempt to understand the situation, hand-in-hand with a creative process where new understandings are manifested and materialized.[3]

Design as Reflection-in-Action

Once interest in Herbert Simon waned in the 1970s, another theorist whom design research came to rely on was Donald Schön (1983). He was interested in reflection as part of the practice of different professions, building on Dewey's ideas to develop his theory of reflective practitioners. Schön argued that seeking to understand is typically what professionals do when they face uncertainty—if instead of focusing on analytic problem-solving we acknowledge that in situations of uncertainty and complexity people reflect, we gain a better understanding of how people tackle things that are surprising and problematic. He particularly drew attention to the drawing and sketching done by architects and designers as a striking example of reflection going hand-in-hand with innovation—in developing new possibilities built on a better understanding of the situation at hand. Schön thus saw sketching as a necessary tool for designers and architects in getting to grips with new and often vague and ambiguous situations or projects with a high degree of uncertainty and complexity—wicked problems.

Schön argued that sketching enabled the architect (and the designer) to relate simultaneously to the "detail and the whole". Like a "zoom camera" (Johansson, 2006) one may in sketching quickly switch from detail to whole and back again, but also between reflecting on the situation and attempting to relate to it through various suggested solutions. Schön considered that in this process the problem in itself is framed in tandem with trying out possible solutions—that, as a process, it is more about understanding or identifying the problem, or "problem-setting", than about problem-solving—even though problems may be solved by it. In other words, the problem, if it exists at all, is understood only at the end of the process, by which time one tends to also have the solution clear in one's head, as well as a deeper understanding of the situation as such.

In innovation research, the design process is often seen as a combination of sometimes converging and sometimes diverging phases (e.g., Howard et al., 2008). However, as a designer, it is not always possible to distinguish between these phases, particularly not at the sketching stage. Here the divergent, in other words the exploring of the situation through different solution suggestions which one can interpret and reflect over, is combined with converging steps, for example critical prioritisations and decisions. Schön expressed this sketching-oriented process as a "reflective conversation with the situation". In this conversation she or he uses a repertoire of impressions, skills, perspectives and emotions, etc. We believe that the designer's experience of creating meaning in a product context in fact benefits all types of innovation, including concrete problem-solving. The design approach expands the scope of opportunity, for example, by challenging the framing of the problem at an early stage of the process.

Design-Driven Innovation and Meaning-Making

Another important contribution to an expanded view of the innovation process is actively incorporating the perspective of the user, an element that is utterly central to the designer. It is easy to consider the more practical aspects of use, such as ergonomics and practical function, aspects that go without saying for any designer or product developer. However, for the designer, use is also about features that are more difficult to encapsulate. Design theorist Klaus Krippendorff (2006) considers that users are not primarily interested in the functional characteristics of products but more in how the products speak to them on a more intuitive and emotional level—that is, what the product means to the user. In order to understand the user's preferences, one must thus also attempt to understand what the user considers important beyond the more functional and technical aspects (which of course are also important). What moves the user? Who does the user want to identify with? What should the experience of using the product be like? For Krippendorff, design is primarily about creating a meaningful relationship between the user and the object:

> The etymology of design goes back to the Latin de + signare and means making something, distinguishing it by a sign, giving it significance, designating its relation to other things, owners, users or gods. Based on this original meaning, one could say: design is making sense (of things). (Krippendorff, 1989, p. 9)

Italian innovation researcher Roberto Verganti (2009) builds further on Krippendorff's understanding, considering that creating meaning is a valid and important area also for innovation. He coined the expression "design-driven innovation of meaning" and argues that the designer is ideally placed to contribute towards such innovation as he or she is trained in interpreting users and society from a socio-cultural perspective, as well as suggesting new meanings. Verganti even claims that one should avoid listening to the user too much. The user can very well contribute towards developing existing products but is simultaneously locked into their own understandings of what the product is about. Sometimes one therefore has to suggest products with new meanings that the users themselves would be unable to imagine without the help of the designer.

While what we have described above primarily characterises the aesthetic and creative part of the designer's professional expertise, there is also a more analytic aspect to design practice—the designer also works with more objective product properties which have to function in a commercial and production-oriented context, such as function adaptation, modularity, cost, etc. Design theorist Håkan Edeholt (2007) considers that the designer is typically part of a "field of tension" between engineering, business and design. However, he also states that in this field of tension the emphasis is on the more concrete and measurable characteristics while the more subjective features and experiences added by the designer are often ignored. It is even more rare for attention to be paid to the designer's professional expertise in its own right. One reason for this may be that aesthetics often come in late in the development process, at least in Sweden and the Anglo-Saxon world where they tend to be seen as a product feature rather than as a professional skill.

In our view, the professional skill of the designer can support the entire innovation process, particularly its early exploratory phase—in what, in the context of innovation, is often termed the fuzzy front end. The innovation challenge taken on may be complex, vague and ambiguous, but taking interpretation, reflection and the creation of meaning as a starting point in the innovation process is anything but fuzzy, at least in design. We thus consider that where there is a desire to innovate from an expanded innovation perspective, the designer's professional expertise is essential.

We will now provide examples of what this perspective may mean, by drawing from an experimental research project in which designers introduced design practice hands-on to 'non-designerly firms' with the aim to boost innovation. We will also discuss economic and managerial implications of such an introduction.

THE EXPERIMENTAL RESEARCH PROJECT

In order to explore how the contribution of design practice to innovation can be understood we set up an experimental research project. We relied on the experience of the Swedish government and SVID's (SVID–the Swedish Industrial Design Foundation) major design initiative in 2003–2005, which had "Design as a Development Force" (Johansson, 2006) as one focus, as a baseline when designing the study. This initiative studied companies that had successfully integrated design in their operations. However, in these cases the actual encounter between design and organizations had already taken place—the studies were carried out after the fact, so to speak. In order to gain a deeper understanding of what actually happens when designers and organizations with little or no previous design experience first come together, we chose an experimental[4] approach that gave us the opportunity to also be on the spot when this happens.

The project, a joint project between SVID and Business & Design Lab,[5] was thus based on staging, studying and attempting to understand what happens when designers actively share their professional expertise with a company that wants to strengthen its innovation ability. The idea was for this expertise to be shared through a series of workshops over the course of about eighteen months per company. The series of workshops was organised in partnership between a selected designer and a company, under the leadership of SVID's consultant and the researcher. During these workshops and other activities a group of staff in the company was able to experience design as professional expertise hands-on, while developing actual product concepts. To create as challenging a situation as possible, we selected companies with little or no previous experience of working with designers. Those involved from the respective companies were a selected group representing the company's development function (with representatives from technology, marketing, production, etc.). The selected designer was specifically matched to each company and we sought a wide range of design expertise. Between them the designers involved represented fashion design, industrial design, textile design, interior design and design strategy.

Five different companies were originally involved in the project, but as two companies left the project, one because of the financial crisis and one because of restructuring in the group it belonged to, we here concentrate on the three that completed the processes: the global technology company Alfa Laval, workwear manufacturer Tranemo Workwear and shower manufacturer Macro International.

EXAMPLES FROM THE COMPANIES

Below we summarise some of the experiences and observations from the project under three different themes—examples of what happened in the companies in their encounters with the designers.

Less Focus on Problems and a More Interpretive Process at Alfa Laval

A designer has to be able to integrate technical, functional and aesthetic characteristics. Here the Alfa Laval case provides an example of what this kind of integration can mean in practice.

In the Alfa Laval case, the designer contributed by working with the project leader to develop an internal process for innovation in early stages—the COIN-process (COIN, Customer Oriented IdeatioN). Development of the process had been commissioned by the company's top management level to boost innovation at Alfa Laval. The process built on intense two or three-day workshops attended by handpicked participants from different parts of Alfa Laval. At the time we started discussing a possible design perspective on the COIN-process a pilot workshop had already been held. The results were promising and project leader Klas Bertilsson was entrusted with working further on developing the process. Before the next iteration of the COIN-process we brought in industrial designer Patrik Westerlund.

The background to this specific workshop was that a shift in technology was under way in the industry. The question was how Alfa Laval would strategically respond to this shift. The unit affected, whose head office is in Italy, decided to try the COIN-process as a way of investigating the challenge. Experienced engineers from recently incorporated production units in the Netherlands, Finland and the United Kingdom were invited in to work with their Italian colleagues to explore the challenge in greater depth in two workshops.

Expanding the Process by Introducing Associative Language

At the first workshop Patrik concentrated on observing the process in action. He did this by participating in the group exercises like any other participant. One of the conclusions he drew was that the process was "too linear"—it lacked more divergent elements in which assumptions and understandings could be explored and examined more freely. He basically found it was the exploration of the customer's perceptions that was not addressed in any depth. Coming from a design perspective, Patrik argued that the user's impression is so central that one might as well start "at that end" rather than with the more technically-focused, problem-solving aspects.

Patrik proposed that the creativity method Synectics (Gordon, 1961) could be appropriate for identifying what might be good customer perceptions to aim for. Synectics is based on a group using chains of associative and metaphorical language to interpret and so approach fundamental assumptions about a particular predefined problem; in this way alternative solutions can also be developed. This is an exercise in which a group can find itself "far out" and which requires an experienced leader to enable the group to finally find its way back, sum up, and formulate potential concrete solutions for the original problem. In order to address the user perspective (rather than a specific problem) Patrik thus contributed with a reverse logic based on defining potential user perceptions of a product or service and *then* determining which solutions would be required of the product or service to achieve these perceptions. The relevance of this was expressed by Klas:

> It became a systematic way of making sure that it really did all hang together every step of the way. Otherwise it's easy to get bogged down in how to solve particular issues and then you get fixated on those solutions. Now we can go back and link it up all the way to the customer perception we want to achieve.

We could express this as the COIN process being developed such that it contained two logics coming together—traditional problem-solving and meaning-making. This meeting of logics offered a different perspective on how the new technology/product could be related to and exploited—ways which competitors applying the new technology more traditionally had not come up with. Here the designer's user and meaning-oriented approach and the engineers' problem-solving approach went hand-in hand in a beneficial way. However, in order to make the meeting possible in the practical aspects of the work, another element suggested by Patrik also played an important and surprising role.

Sketching Contributing to Conversation and Reflection
As discussed in the introduction to the chapter, in design practice sketching is a crucial tool for dealing with complex, qualitative, and sometimes conflicting characteristics and perspectives in the creative process. Patrik proposed a 'sketcher in residence' and hired design student Adam Henriksson for the job. Adam was placed at a table in the middle of the conference room with pencils and paper while the four groups worked at their tables around him. The idea was that Adam would be on hand to illustrate and visualize ideas from the groups. But as it turned out this sketching instead contributed to a shared reflective process between Adam and the groups.

In the process, typically a representative from a working group would go to Adam and start to explain a principle to Adam, such as an idea for a unit constructed differently, which he wanted visualised. During the sketching process and in the conversation with Adam, who attempted to interpret the idea by asking questions that were rather naïve—as Adam didn't know anything about the technology, the representative from the working group experienced 'light bulb moments' of sudden insight when he had to consider Adam's questions, insights that challenged the original idea and helped to refine it further. When the visualisation was then shown to the rest of the group, additional new ideas emerged which were then taken back to Adam. An unexpected and exciting dynamic arose through 'collective sketching', or as expressed by Patrik:

> Adam was forced to characterise what the others said by capturing what they said, and naturally he misunderstood things a bit all the time. What on earth are they thinking? And that misunderstanding produced a discussion and could result in a new shape, a new function or a new solution.

As a result of these new elements, the focus changed in the second workshop and the process was partly liberated from its previous one-sided focus on technology. Instead, the workshop more clearly came to be about the possible benefit and the importance of the new technology in question for customers and users in the future—in other words what had been sought from the very beginning. What could the new technology, which gave the product completely different dimensions and geometry, offer in terms of new opportunities for customer benefit? It was not until a later stage, once the discussion had expanded to talking about the meaning of the technology in general and its benefit and value for the customer from a user-oriented perspective, that the debate linked back to the more technically focussed discussion, partly through the sketching process. This time, however, the discussion had gained new perspectives and understandings. The result was a consequence of an expanded perspective based on the likely needs and desired perceptions of the user, which partly led to developing new maintenance opportunities, placement alternatives, modular options and aspects concerning the unit's appearance (for example placing it in a highly visible spot as an architectonic element rather than placing it on the rooftop). In other words, the development of the design went hand-in-hand with the development of the technology in an in-depth process that demanded that the design perspective was involved from the very start. The relevance of these and other contributions that Patrik made to the COIN-process were summarized by Klas:

> We are quite good at solving problems at Alfa Laval. But it is more rare that we fundamentally challenge the very problem that we attempt to solve. Approaches for doing this is what design has contributed with.

More Complex User Dimensions Integrated in Tranemo Workwear

Taking an active interest in the customer is of course fundamental to all parts of an organization. Nevertheless there are clear differences in the approach to the customer—or "the user" and "the user situation" as the designer usually puts it. Designers take the user as their starting point in several specific ways from which other groups in a company can benefit. The following story shows how the designer contributed towards an entirely new understanding of the user or customer.

Tranemo Workwear has been manufacturing clothing for over 75 years. Today the company is one of the leading companies in Sweden—especially in workwear for heavy industry. However, the company had never considered that it needed a designer, believing that workwear looks the way it does as a matter of custom and tradition. Nevertheless, it had been noticed that a younger generation of customers were increasingly also seeing workwear as a marker of identity. The company had also realised that their clothes are mainly designed for men and that there can be a problem finding good clothes for women in male-dominated industries.

When we introduced the idea of a design project for the company, the management saw in this an opportunity to work on new product development strategies. Fashion designer Charlotta Schill, selected by the researchers, was unusual for also having had experience in workwear design. This was a major plus—when she challenged the group, they trusted her judgment because of her workwear experience, but the fashion perspective also made it possible for her to challenge the group in different ways. Here we concentrate on two of the workshops that focused on studying the user in various ways.

Disturbing Tranemo's Established Understanding of Work Wear

At an introductory workshop the aim was to "shake up" the company's view of workwear. Fashion design students at the Swedish School of Textiles at the University of Borås were invited to spend a day studying Tranemo Workwear's clothes and to challenge them from a fashion perspective. The results of the workshop comprised everything from sketches changing the look of existing garments to completely new types of workwear. Several of the suggestions directly impacted on the next collection of garments for tradespeople, including choice of materials, colours and cut. For example, the classic 45 degree-angled corners on the pockets were rounded off—a stylish innovation which several of the students insisted on that also had practical benefits, for example, so nails could not get lodged in corners.

In the process that followed the initial workshop, the early workshops were about freely but tangibly searching for inspiration for a theme that would be interesting to work on further. The participants composed so called 'mood boards'[6] to express emotions and aesthetic impressions (styles) by cutting inspiring pictures out of newspapers and fashion magazines. They also searched the Internet for current trends in other related sectors (e.g., sportswear) and subcultures (e.g., hip hop). The results of the workshops led to three main themes being identified for further work: (1) workwear for women in male-dominated occupations/industries; (2) workwear for a younger generation; and (3) more ecological workwear.

Walking In the Shoes of the User

In a later stage of the project Charlotta took the initiative to visit one of Tranemo Workwear's most important buyers, the SSAB steelworks in Luleå. She argued that it was important for the development team at Tranemo to gain a 'front row view' of the everyday life and environment of the wearers of their garments—to *experience* the heat of the liquid metal in the smelting works, the thick layer of dust on every surface, the boiling hot windows of the monitoring rooms and the cold air from the air conditioning. This was something none of them had experienced before, but something which goes without saying for a designer—the necessity of experiencing the user's situation first hand. The visit provided a wealth of insights into the garments and their design, but also some unforeseen effects. For example, conversations with the female employees revealed that they were unaware that flameproof underwear was even available—a serious deficiency and an opportunity for Tranemo's sales department to communicate the fact that the company actually sells these garments.

In conjunction with the study visit a workshop was also arranged jointly with Luleå University of Technology (LTU) within the framework of their research project The Factory of the Future. As well as being attended by two representatives from Tranemo, the workshop also involved a group of workers from SSAB and researchers from LTU. Together they explored the potential workplace of the future in heavy industry and used different brainstorming techniques to come up with ideas and visions for the work wear of the future, as well as forging relationships and a desire to continue working together in the future.

Back at home, work continued in different workshops until a range of conceptual garments for women in industry had been developed, garments that in different ways challenged taken-for-granted assumptions about work wear—for example, challenging that they had to be made out of heavy textile to be protective with a concept that used highly protective reinforced stretch nylon that made the garments radically more comfortable.

Designer Charlotta's starting point, as is characteristic of the Scandinavian design tradition was the user and the user's situation—the designer seeking *direct contact* with the user rather than using sales figures, questionnaires, or the opinions of salespeople. What we described above were three different ways of getting closer to the user: (1) working with trends, identities, and other channels represented by media; (2) experiencing the environment in which the products are going to be used—which includes gaining a *sense* of the environment and how it is *experienced* by the users; and (3) presenting techniques for 'co-designing' which actively involve the user in generating ideas and being part of the development process. All these ways of working and the processes that the designer introduced were new to the company, and also broadened the understanding of what design could offer, or as expressed by CEO Max Larsson:

> You could put it like this...Design has always been questioned—what is it they do? All you need to do is make a pattern and it will be fine. But there's so much more to it than that. It's obvious, if you get this sort of input from outside then you can probably put much clearer arguments to the board, argue that we really need to invest in the design department.

Cultural and Aesthetic Dimensions Incorporated in Macro International's Development

Most occupations are well aware of the importance of connecting with the outside world. Designers often incorporate the outside world in a broad perspective in which cultural and aesthetic dimensions are of major significance. Possibly what is crucial to this way of monitoring the world around them is the way in which the designer, in gathering different types of both factual and inspirational material, circles around the question they have been set to solve or the commission they have been asked to complete. Monitoring and gathering facts and inspiration involves widespread scanning, covering everything from materials, technology and trends to business concepts, social development, etc. In seeking a context and starting points for the questions posed by the issue at hand, the logic tends not to be direct or comprehensible in a linear fashion. Perhaps it can better be described as a platform of starting points for the designer. This then becomes a way of expanding the companies' view of the development task, and possibly also a good starting point for business strategies. The way in which this worked was shown particularly clearly in the Macro International case.

Widening the Area of Interest and Inspiration

Macro International is a typical small industrial company set up 25 years ago by an entrepreneur who identified a gap in the market for bathroom products—the company is based in southern Sweden and develops and manufactures showers. Macro International has been part of the Norwegian Tema Group since 2002. Designer and design strategist Cecilia Nilsson was brought in to work with the development team. After introductory workshops she noticed that there was no detailed understanding of the product itself—it "floated freely in relation to the world in which it found itself". When she also noticed that the business concept did not appear to be sufficiently well-anchored she decided to start where many designers themselves start—in the outside world. To develop the company's awareness of the wider sphere in which they were operating, she chose to work with the company to plan and carry out a number of activities which had never been done before. For example, they attended a technology fair in Germany and an interior design fair in Italy and visited art exhibitions in Denmark and Sweden. A workshop was also held on trend spotting in the media. The point was not just to obtain information but was just as much about searching for *inspiration*.

Towards a New Product Understanding as a Foundation for Design

In the next stage the group revised the company's sales catalogue that felt anonymous and dated. During the discussions with the designer, the group reached the conclusion that the company should express itself differently in terms of graphics, images and text. Being forced to create an identity for the company in the catalogue also led them to start thinking through and articulating a broader and clearer understanding of Macro International's product—which in turn came to be encapsulated in a new company slogan—Your moment. This clearer view of the product became part of work on the identity of the products and the company itself, or as said by marketing manager Magnus Reinhold:

> It was definitely inspiring and an experience and we talked about a theme that we call your moment. Wellbeing. Suddenly the shower is about wellbeing,

having a moment to yourself to enjoy and relax. A shower suddenly took on a whole new meaning.

That done, the product development work could be tackled, now with a clear focus serving as inspiration for new ideas. At subsequent workshops several concepts were developed, particularly technical ones that were immediately related to the new understanding of the shower. At these workshops a great deal of sketching also went on, in two and three dimensions, in scale and full-sized models, something which particularly challenged the more traditional design work in several ways, as expressed by engineer Daniel Jacobsson:

> Yes this was one of the last things we did, building full-scale models. It's a way of working I've never used before. We've made prototypes but then you have to mill out bathtubs, source glass, produce panels... A very long time to produce a prototype in other words. And what we did here took three hours and we had a full-scale prototype in front of us. It's something I'll use just to see the shape and yes the feeling of standing in the shower. How big it's going to be. How easy to fit it is. You get lots of answers and you can see a whole lot even with what is a very simple rough model.

A Rewarding Partnership—Lessons Learned and Economic implications

In the study we concentrated on relatively traditional companies with no or limited experience of design. Not all of the companies underwent the same process of change—here we are not presenting a one-size-fits-all model for how to work with design. Nor was this our aim. We have discovered that in the first instance it is not about constructing a number of design methods and specific approaches which can be generalised. Instead we have come to see it as a question of passing on the designer's professional expertise by emphasising the experience of design as "doing"—learning by doing. Professional expertise in practice is much more complex than can be described in a method or a process and we believe that such a perspective also provides greater success in concrete terms. However this demands a different view of "implementation" than is common in a consultancy context.

None of the companies could themselves have achieved what the designer added—all the companies agreed on that. Despite the fact that they did receive something that afterwards they said they were missing, integrating design work in the companies was far from a simple process. It took input from SVID and the researcher to help with the integration work and to support the designer who took on an unfamiliar role in the project as a coach or mentor. This is interesting and worthy of further thought. The link between the designer and the company is a 'matchmaking' process that demands sensitivity—an area in which SVID has great experience and which was an important factor in the majority of the projects with which SVID has been involved. Despite this, "our" companies initially found it hard to see how they could use a designer in the ways we described and it therefore took committed key staff in each company with inquiring minds to persist to achieve the changes to internal working practices that proved to be necessary.

The designers' input took on several different characteristics are the result of the project's open approach. No company became such self-starters that they were able to manage without a designer (and nor was that an aim). In some cases the designer introduced working methods which the companies themselves have been able to adopt, but above all the project seems to have led to the companies becoming better at hiring designers, or employing a designer because they have realised how design can benefit them at an early stage of the development process.

Here we present four lessons learned and, in conclusion, examples of how the different companies view the potential results of the design initiative in economic terms.

Lessons Learned

The foremost experiences—and we believe lessons for other companies—are:

Lesson 1: Expanding The Boundaries of What Is Taken For Granted

Reflection and interpretation are not something unique to designers but something that arises in all professions. However, for the designer, reflection and interpretation are essential tools for creating meaning and developing product concepts with new meanings. We might say that the designer progresses from a question to a finished product. It is through querying and by asking questions that the designer interprets complex contexts which encompass non-objective features such as aesthetics, emotion, and the user's values. The nature of the designer's interpreting and questioning is something that everyone in the companies in the project was able to experience. By questioning the things that are taken for granted in the organisations, the understanding of the product being developed and manufactured was extended and expanded. After the project, a shower in the Macro International case, for example, or a garment as in the Tranemo Workwear case, were partly seen in new ways. We do not know how deeply this questioning attitude became embedded in the organisations. Interpretation must be a constantly on-going process if innovativeness is to be strengthened. In interviews, however, several employees have stated that they now examine the products and the outside world more critically and more consciously.

Lesson 2: Sketching and Visualisation as Central Tools

Images and active visualisation are important tools in the interpretive and creative design process. However, in the companies the images were not primarily intended to illustrate and communicate ideas, a common preconception when it comes to design, but were central to the actual creation of ideas. The "sketching" process took varied forms, starting with a pencil and using materials. In the Alfa Laval case we saw sketching become a social process. Sketching was also a way of approaching and identifying trends and tendencies which are harder to articulate. Working with images also made people more aware of the things that are difficult to express, which cannot be pinned down by language. In all the companies people were initially unfamiliar and possibly also uncomfortable with visualisation and sketching—that it felt like being at children's day-care. But this perhaps says more about the value we place on creativity in society in general. During the course of the project, all the companies came to feel more comfortable with sketching out ideas in various ways in pencil or using materials, in two or three dimensions, and with working with images. Although at the same time we saw that the companies have far from adopted the designer's way of thinking by sketching an important barrier has been broken down.

Lesson 3: A More Complex and In-Depth User Understanding

The user is always the focus of the designer's interpretive process. The designer takes great pains to build up as complex as possible an understanding of the user and the situation in which the product is used. We have seen several examples of the designer virtually dragging the staff out of the office and into the field and also of how meeting the users immediately resulted in new thoughts and ideas. One element in this deeper understanding is also studying prevailing trends and the socio-cultural climate—taking the contemporary pulse. Such processes may be hard to justify at first glance, for example the benefit of a visit to an art gallery, a steelworks or a university. Once this fear was overcome, however, it was striking to see how surprisingly quickly such activities resulted in new angles. However, there remains an unfamiliarity with sensing what suitable activities might be, and here the designers were and are active at giving advice.

Lesson 4: A Broader Perspective on Innovation

Through the designer's input the innovation process has come to be about features extending beyond the traditional technical understanding. The most radical element is an understanding of how the meaning of a product created by the user is also a potential area for innovation—how, for example, clothes are socially understood rather than representing fixed norms. In other words, how a company can influence and address these on-going and changeable constructs by actively suggesting new meanings. Interestingly enough, these processes also force one into solving technical problems, and it appears to be precisely this *combination* of technical expertise and understanding of the creation of meaning that characterises potential new innovation areas for the companies.

Economic Implications

But what are the companies' economic arguments? The costs are fairly simple to calculate but it is harder to say anything unequivocal about the gains to be made. Design is seen as an investment—but an investment that is often conflated with others. The value gained is often indirect and hard to measure but because the returns are said to be large in relation to the inputs, it is worth reasoning out the nature of the return and estimating the magnitude of the value gained.

Tranemo Workwear
CEO Max Larsson says that as far as Tranemo Workwear is concerned, the project, together with other initiatives, has led to gaining new perspectives on opportunities in the industry, increased the company's margins and established a more consistent product development process. The financial contribution can be linked to how in the past four years Tranemo Workwear has reduced its sales of 'basic garments' by roughly 50% as a consequence of an active focus on the more specific areas of flameproofing, high visibility and identity. During the same period, sales of a number of 'design garments' have increased by almost 400%. Although in 2010 the number of these garments sold is as yet only half the number of basic garments sold, the contribution margin for these garments is almost three times as high. This means that thanks to the new strategy the reduced sales of basic garments are more than offset by fewer design garments. No one believes that the design input was the only cause of the increased contribution margin and the increased number of design garments sold but if it only contributed let us say a fifth, it would still be a profitable project even in the short term.

Macro International
Macro International manufactures almost twice as many showers as its nearest competitor, which is a bigger household name. One of Macro International's challenges has been to strengthen brand recognition, which according to Macro International's CEO Mikael Lunneryd stands at about

20–30%, compared with about 90% for its competitor. Mikael thinks that if within three or four years the company can gain greater recognition to the tune of 20,000 showers, so increasing the margin per shower sold by around 10 to 15 dollars or so, the money invested will more than have paid for itself.

Alfa Laval

In the Alfa Laval case the argument is a different one. The project has contributed to an important internal workshop-based concept development process. Project leader Klas Bertilsson estimates the costs of each such workshop initiative as some thirty- to forty-thousand dollars. It is vital that the process is of such high quality that it succeeds; otherwise there is a major risk either that the company will lose its edge over its competitors or the launch of new products will be delayed. On the other hand, if this kind of process succeeds in developing a successful concept, he says that the income for the company in the longer term may be in the region of tens of millions of dollars. If the design perspective has helped to safeguard the value of the process, the amount brought in by the project may be huge compared with a cost that is modest in this context.

MANAGERIAL IMPLICATIONS
OF DESIGN-DRIVEN INNOVATION

It may seem that introducing design was smooth sailing. But there were a number of challenges, not least managerial ones—challenges that we think indicate the need for new managerial practices in this millennium, when more traditional analytic approaches are complemented by more artistic and interpretative ways of engaging with and understanding the world, the market place and one's products.

The single most important decision when we established the approach for the processes was to privilege the *practice* of design. This meant insisting on processes based on the experience and knowledge of the involved professional designers, and making certain that these processes would not get caught up in or become too influenced by established processes and approaches in the companies, not least traditionally analytic and problem-solving oriented ones (e.g., engineering and management). This meant that at least six management capabilities needed to be developed. In the cases the management capabilities, including knowledge, responsibilities, actions and decisions, were actually split between the managers and the designers. But to flesh out recommendations we will assume that these are all management challenges.

Management Capabilities

Management Capability 1: Acknowledging Organizational Situatedness

To be able to develop innovative concepts that do not fit within the existing framework of understanding in an organization it is necessary to develop *new* understandings. This entails the need to develop processes for actively exploring *new* meanings, for example what a product *could* be about or how a novel technology *could* be understood. This also means that it is necessary to challenge the status quo, the current 'meaning-space' in the organization to develop a foundation for imagining and materializing something new.

This need for new understandings has a number of management implications. First, it means establishing the acceptance that the organization is inevitably 'situated' in tradition, conventions, culture and so on, and accepting that there is no possibility of suddenly "thinking outside of the box." In relation to this, it is also necessary to establish an awareness of how established pre-understandings in the organization, for example social dimensions such as norms and values as well as technological and embodied preferences, are typically unreflected and taken for granted as truths, and are typically defended, reproduced and institutionalized.

Second, it means building acceptance of the fact that it is actually possible to understand *differently*, and to activate an ambition in the organization to seek such different understanding. This is fundamental in preparing the organization for active meaning-making that will necessarily disrupt the status quo, and has radical implications for management as conflicts and issues of power cannot be avoided. Instead it is necessary to develop an understanding of how friction and conflicts are necessary and welcome dimensions of the process.

Third, it means to develop an understanding of how situatedness, to be placed in a context, often discussed as 'prejudice', is not necessarily bad, but actually constitutes the fundamental resource in meaning-making. By being situated, by already being immersed in understandings, the organization has access to the means necessary to also *expand* the 'organizational horizon of understanding', and open up to *other* ways of understanding as fundamental to innovation.

Management Capability 2: Activating Organizational Meaning-Making

An overarching capability is the ability to instrumentally *activate* 'organizational meaning-making' to enable innovation. As a management challenge this is inherently difficult for several reasons. The first three ones are described above. In addition to these, activation of meaning-making also has to work against the fact that company organizations are often steeped in and organized according to a functionalist school of thought, with dominant perspectives (such as problem-solving) and knowledge ideals (for example emphasis on verbal concepts, objectivity and cognition) resisting the interpretative concept of meaning-making. This implies at least two major management challenges.

First, activating meaning-making requires bringing on-board a broadened perspective on knowledge and introducing an interpretative and constructivist framework in otherwise functionalist oriented organizations. Unfortunately, most functionalist oriented organizations lack concepts and vocabulary for articulating, and thus understanding, or in any nuanced way deliberating the implications of this requirement, let alone for being able to develop, operationalize and actively exploit meaning-making oriented processes and practices (Lester et al., 1998). One consequence is that management has to be able to actively develop knowledge about meaning-making and understand its relevance for innovation, and also be able to introduce the necessary concepts in the organizations, and legitimize these and what they entail for the organization.

Second, management has to actively *resist* attempts at subsuming meaning-making related innovation processes under functionalist processes and procedures (such as the linear stage-gate procedure and typical management process in general) as this inevitably means the destruction of the logic of meaning-making. Instead managers have to define specific organizational space, requisites and resources for enabling innovation processes related to meaning-making, *alongside* functionalist processes. This also includes the organization of interaction between such processes.

Management Capability 3: The Management of Conversation

In processes of meaning-making the new emerges in the 'in-between', as a "conversation with the situation at hand" (Schön, p. 43). This means that when meaning-making is activated in groups the new will emerge in the intersection between the members of the group and the conversation with the situation (a market challenge, a novel technology, a user situation etc.) that is interpreted. Ensuring that the new emerges *between* the members of the group holds several important management challenges.

First, when activating meaning-making in organizations management's ability to socially engage groups in conversational processes of meaning-making comes into focus. In such processes management's ability to encourage trust, sharing and openness is of critical importance, but also the ability to emphasize a fundamental ambition to seek understanding *together.* Managers need to establish the social as the space where new understanding and innovative concepts emerge *between* the members of the social situation. Here individuals and specific ideas become less important than is the norm in innovation management. Instead *depth* and *understanding* through conversation takes precedence over ideation-oriented attempts at building innovation on fast iterations of specific ideas, even though such approaches may also be useful *within* the process of meaning-making.

Second, in meaning-making complex and composite concepts emerge in tandem with the ability to pose relevant questions, including ones that are critical, phenomenological, or naïve. Here managers need to resist an urge to request an immediate answer or solution, but instead pose questions that further *understanding*, for example of what a product is *about*, even when these questions are critical to the status quo. This also means that managers have to see the interest of the individuals and group as central in the process of meaning-making, even though this interest at the outset may only seem vaguely related to a specific problem or issue.

Third, when managing the social as the space for innovation, managing itself takes on a different character than is the norm. When each individual contributes on equal terms, and when domination of a single person, for example a manager, risks spoiling the process, 'facilitatorship' becomes more important than traditional leadership. Important dimensions of the capability to manage the social as the space for innovation include the ability to organize such space as on-going (rather than project specific) conversations across organizational boundaries and professions; the securing of necessary resources, for example to develop a physical space for meaning-making oriented innovation work, and to permit expenses that may seem unusual (for example trips to art shows and unusual subscriptions).

Management Capability 4: The Management of Emergence

As described in the above capability, meaning-making can be seen as a conversation with that which one wishes to interpret and understand. Such processes are gradual and *emergent*, and understanding, and the accompanying result, comes late in the process. Further, understanding and the result emerge as *complete* concepts, as understanding *materialized* in its entirety. This emphasis on completeness, that is more akin to artistic creation than analytic problem-solving, is in at least three respects at odds with reductivist approaches common in functionalist oriented organizations.

First, reductivism builds on the notion that something can be divided into its component parts. This does not work for experiential, sensory and meaning-oriented issues, even though in the process critical and reductivist devices may also be necessary, as long as they are not allowed to define the process as such. Second, reductivism demands that the problem definition is made "up front" to enable a reduction of the problem into its composite parts to be solved or carried out—that is, it starts with and does not challenge an established understanding. This makes further meaning-making impossible. A readiness to challenge the problem is thus crucial at the outset. Third, and related to this, reductivist approaches builds on a temporality opposite to meaning-making. Instead of clarity at the outset, clarity, and the result in, comes late in the process.

To managers more accustomed to analytic and reductivist approaches this means having to more or less completely revise how a solution or result is arrived at, implying three fundamental challenges. First, it means developing the capability to build organizational trust in the different temporality of meaning-making and actively avoid the organizational temptation to require up-front problem definitions or demand premature closure. Second, it requires the capacity to manage the necessary phase of 'being in the open', of being in the situation where established understandings have been challenged and begun to dissolve, but before new understanding has solidified. Third, it requires the ability to instil trust also within the group, and to develop different measures for keeping the process in motion.

It is also important to note that this emphasis on a meaning-making oriented temporality means to challenge hopes of achieving effects through the implementation of certain tools or quick workshops within *existing* processes as being too naïve. Meaning-making takes time and has to be seen as on-going process with its own character. It is a process that has to be constantly tended and renewed, as innovation is a moving target. Ultimately the management challenge is to enable this type of process alongside, and interlinked with other processes without the unique character of the process risk being ruined.

Management Capability 5: The Management of Making in Innovation

Meaning-making does not have to be restricted to verbal conversation, quite the contrary. In design, with its roots in crafts, conversation is always entwined with *making*. When innovative concepts are compositions of functional, semantic, aesthetic and experience-oriented dimensions, rather than being restricted to what can be expressed through language alone, it becomes crucial to engage in meaning-making that also involves the senses and embodied knowledge. This makes it possible to connect the process of innovation with all dimensions of how a product or services will be experienced.

Visual and material making also holds the capacity to let organizations release from deeply rooted aesthetic and sensory pre-understandings or preferences that may be impossible to challenge or even capture verbally, and that may stand in the way of innovation. Further, making emphasizes the *thing* (the emerging product or service as understanding materialized) as the central focus of meaning-making: it is the thing that emerges between the individuals involved in meaning-making. And it is the thing that makes this process possible at all—that makes it possible to unite the process across practices, different knowledge traditions and so on. The thing defines the space for innovation and it provides the possibility of 'common ground'.

But making holds several management challenges. First, when engaging in embodied and aesthetic knowledge, organizational issues that are considered uncomfortably subjective and politically sensitive may be revealed, thus requiring the management ability to deal with such issues productively, without repressing them. Second, making involves hands-on practices that most non-designers or non-artists may be unused to, and which may lead to at least initial resistance. Third, making draws on subjectively oriented knowledge that may be at odds with the knowledge perspective in organizations steeped in a functionalist knowledge tradition—for example, knowing what is 'right or wrong' from an aesthetic perspective. Fourth, making is often associated with less 'serious' work. It may feel 'daycare-like' and might be difficult to defend in a more verbally and rationally oriented organization. Fifth, there are often assumptions that the result, for example of sketching, should be artistic and 'nice', which may generate anxiety, especially when a professional designer is present.

Thus the management of making holds several challenges. Not only do these five challenges have to be transcended, they also imply the necessity of creating meaningful and useful situations for making that actively involve aesthetic dimensions. To do this takes artistic knowledge and experience that managers may have difficulties developing, and that may not be readily available in functionalist organizations. Otherwise there is a risk that making never transcends or deepens the process beyond post-it note exercises or prototyping as solution-testing. Instead making relates directly to the process of exploring and interpreting the completeness of a potential experience of a product or service, including both functional and aesthetic dimensions.

Management Capability 6: The Management of Inspiration

To engage in meaning-making demands *immersion* in a wide array of knowledge—the ability to interpret and deliberate knowledge from quite different sources and realms. Some of this will be at odds with favoured knowledge perspectives in many functionalist oriented organizations. It may even involve the application of knowledge that may be seen as 'useless' in the sense of not being immediately tied to utility function, and is not necessarily 'objective' in the sense of being quantifiable or possible to verbalize. Such knowledge, that supports understanding by enabling a holistic perspective, is often called *inspiration*. This includes aesthetic and experiential knowledge. Further, in meaning-making a *surplus* of knowledge is necessary to enable interpretation, deliberation and manifestation of meaning.

For management this means supporting immersion in knowledge that does not necessarily make sense at first, and that may be difficult to defend or motivate to anyone outside of the process of meaning-making. It means to appreciate aesthetic knowledge for the richness that it brings to the process, and the ability to articulate this to others in the organization. It means to understand the relevance of experiencing 'first hand', for example, experience of a user situation, and how it activates a wealth of necessary knowledge, such as sensory and emotional knowledge, that can never be communicated via questionnaires or second-hand reports.

Managers, when guiding the process, also need to develop the ability to sense and suggest possible areas of knowledge in which to immerse, for example, the possible relevance of going to a certain art exhibition or visiting a trade fair far outside the sector of the company. Further, managers must be able to defend the need for inspiration and other types of 'useless' knowledge, including associated costs, in relation to the management of the organization. This also includes that ability to establish strategic collaboration with actors representing knowledge realms outside the norm of the company, for example artists and other actors in the socio-cultural sphere, to have access to emerging meanings.

CONCLUSIONS

The above set of management capabilities indicate an insurmountable host of challenges for most managers if they are expected to operationalize innovation inspired by design practice, especially if they, as most managers, lack artistic training and experience. Despite this, these challenges *were* tackled in the cases presented above. This was achieved through the strategy of "cutting straight to action"—by activating meaning-making in multi-disciplinary groups through hands-on design work. This work was facilitated and guided by artistically trained designers. But even though the managers were not active in the processes, their support was important in legitimizing, defending and securing resources for processes that were foreign to the organizations. This implies the need for close collaboration between managers and designers in all design-driven innovation.

But what is most interesting is that, beyond initial friction and difficulties, the individuals in the different groups engaged productively in processes that could be seen as artistic meaning-making, despite not having previous artistic experience, and beyond that, it seems that aesthetic deliberation gave the process of meaning-making its very momentum—providing the 'dynamics of innovation' its necessary 'fuel' (Godoe, 2011). We argue that the designers activated and legitimized necessary aesthetic and experiential knowledge in the groups that was dormant but not unfamiliar once engaged. The designers also provided complete processes for doing so. The concepts that were the results in the three interventions emerged in the social interaction *between* the individuals in a process of entwined conversation and making. In other words, the designers, by drawing on design practice, tapped into and activated a potential for innovation that has hitherto been little explored and exploited.

NOTES

1. This study has been funded by VINNOVA, the Swedish Governmental Innovation Agency through the Leadership, Creativity and Work Organization programme (LEKA). Correspondence concerning this article should be addressed to Marcus Jahnke, HDK, School of Design and Crafts, University of Gothenburg, Box 100, 405 30 Gothenburg, Sweden. Contact: marcus.jahnke@hdk.gu.se

2. The term "wicked problems" was coined by Rittel and Webber (1973) as a way of describing complex and multi-dimensional problems in social planning which have no "right" answer or optimum solution and which cannot be tackled by purely analytic means—which had been the dominant approach until then, and still often is.

3. In the PhD thesis *Meaning in the Making—Introducing a hermeneutic perspective on design practice to innovation* (Jahnke, 2013), where these thoughts are explored in much more detail, the framework for discussing meaning-making is based on the philosophy of Hermeneutics—a philosophy in the European tradition with many similarities to American Pragmatism. See also the journal article *Revisiting Design as a Hermeneutic Practice: An Investigation of Paul Ricoeur's Critical Hermeneutics* (2012).

4. From a research perspective, the study is based on an interpretative and qualitative (rather than quantitative) approach. This means that, rather than attempting to establish normative or explanatory models or quantifying results, the *understanding* of what takes place in the encounter between the designer and the company is central. The study also deploys a narrative approach that means that we are interested in the stories and events that those involved find meaningful.

5. Business & Design Lab is a partnership between HDK, the School of Design and Crafts and School of Economics and Commercial Law at the University of Gothenburg in the field of design management.

6. Mood-boards are collages mostly made of images but also for example material samples, texts and colour samples. The mood board has several purposes. It can be used as inspiration to capture a certain mood or style to depart from in design work. It can be used to represent the result of a research process. And it can be used as a communication device to discuss issues of style, emotions, moods and so on, for example with clients (e.g., Eckert & Stacey, 2003; Godlewsky, 2008; McDonagh & Storer, 2004).

REFERENCES

Brown, T. (2008, June). *Design thinking*. Harvard Business Review.

Dewey, J. (1934). *Art as experience*. London: Perigee.

Edeholt, H. (2007). *Design and innovation*. In S. Ihlstedt Hjelm (Ed.), *Under Ytan: En antologi om designforskning*. Stockholm: Raster.

Godoe, H. (2011). Innovation theory, Aesthetics, and science of the artificial after Herbert Simon. *Journal of the Knowledge Economy, 3*(4), 372–388.

Gordon, J. J. W. (1961). *Synectics: The development of creative capacity*. New York, NY: Harper & Row Publishers.

Howard, T. J. et al. (2008). Describing the creative design process by the integration of engineering design and cognitive psychology literature. *Design Studies, 29*.

Jahnke, M. (2012). Revisiting design as a hermeneutic practice: An investigation of Paul Ricoeur's critical hermeneutics. *Design Issues, 28*(2), 30–40.

Jahnke, M. (2013). *Meaning in the making: Introducing a hermeneutic perspective on the contribution of design practice to innovation*. University of Gothenburg.

Johansson, U. (2006). *Design som utvecklingskraft: En utvärdering av regeringens design-satsning 2003–2005*. Växjö: Växjö University Press.

Kelley, T. (2001). *The art of innovation: Lessons in creativity from IDEO, America's leading design firm*. New York, NY: Doubleday.

Krippendorff, K. (1989). On the essential contexts of artifacts or on the proposition that "design is making sense (of things)." *Design Issues, 5*(2), 9–39.

Krippendorff, K. (2006). *The semantic turn: A new foundation for design*. Boca Raton, FL: CRC Press.

Lester, R. K., Piore, M. J., & Malek, K. M. (1998, March-April). Interpretative management: What general managers can learn from design. *Harvard Business Review,* 86–96.

Rittel, H. W., & Webber, M. M. (1973). Dilemmas in general theory of planning. *Policy Planning 4,* 155–169.

Schön, A. D. (1983). *The reflective practitioner: How professionals think in action*. London: Basic Books Inc.

Simon, H. (1969). *The sciences of the artificial*. Cambridge, MA: MIT Press.

Verganti, R. (2009). *Design-driven innovation: Changing the rules of competition by radically innovating what things mean*. Boston, MA: Harvard Business Press.

CHAPTER 10

THE TIMES THEY ARE A-CHANGIN'... ARE WE?

Frame-Changing Design for Military Interventions

Ben Zweibelson
Grant Martin
Chris Paparone
U.S. Army

ABSTRACT

These three military authors offer a unique perspective into design thinking and military sense-making. They offer an argument that while the military as an institution prefers to make sense of the world through a systems-analytic paradigm that is increasingly problematic, a frame-changing design approach is needed instead. This approach requires a greater appreciation of uncertainty and the awareness that new knowledge production must emerge from beyond the single preferred military paradigm.

Millennial Spring: Designing the Future of Organizations, pages 191–224
Copyright © 2014 by Information Age Publishing
All rights of reproduction in any form reserved.

The world as we have created it is a process of our thinking. It cannot be changed without changing our thinking.

—Albert Einstein

INTRODUCTION

The American military today is at a crossroads. After more than a decade involved in counterinsurgency operations and defeating terrorists, it is faced both with the dire fiscal reality of the current U.S. federal budget as well as the war-weariness of the American people. Complicating matters is a feeling by many that Americans are strategically adrift—that there really is nothing guiding United States foreign policy through what appears to be a very complex, new world. This unfolding world reflects a perceived disorder since the fall of the Soviet Union and the terror strikes in New York and Washington D.C. in 2001. The 'rules' of war are constantly changing; the players are as well, and it seems increasingly more difficult to make sense of it all. Many political pundits as well as those with direct military experience have suggested that new ways of appreciating the complexities of military involvement are needed. The world of conflict is neither as 'regular' as we used to imagine it to be nor would like it to be. Better stated, the military science we are comfortable with is inadequate in helping to frame our actions in "irregular," constantly morphing forms of military intervention.

The change in our social world also seems to be speeding up as we become more interconnected and interactive. Technology, globalization, and the rapid exchange of ideas have solved many of yesterday's military problems, only to generate newer, bigger ones that appear to resist our doctrinal methods that we thought would deliver a sense of victory—it's not over "till it's over, over there." For those policy- and strategy-makers who attempt to define actions through the traditional mindset of the dominant rational-analytic frame, as was the case with operations in Iraq and Afghanistan, the outcomes will often seem confusing and contribute to a lack of satisfaction and for a World War II-like decisive victory. Our enemies morph and seldom organize in recognizable patterns, and although our weapons have become far more precise and powerful, they do not have the decisive impacts that we thought they would. Results of traditional systems analysis applied to how targets will be engaged are executed with deliberate acts of force and destruction. Such traditional planning methods provide the illusion of success; however, they tend to run us adrift in a sea of strategic confusion. This confusion will not get better with improving systems-analytic techniques.

On the contrary, 'finding' systems regularity in an irregular, complex, and chaotic situation distorts reality, creating a false sense of understanding.

We, the authors of this chapter, like to think of ourselves as "muddy boots observers," with over 70 years of combined military experience in interventions spanning Iraq, Afghanistan, the Balkans, and the Americas. We propose that military practitioners first need to reflect critically on how the institution tends to frame novel situations in habitual ways and ask, "Why are we framing this way and what are other ways to frame?" Ben is a career Infantry officer and graduate of the U.S. Army's School of Advanced Military Studies (SAMS), the advanced planning school for the military. He also received his undergraduate degree in graphic design. Grant was selected into the U.S. Army Special Forces (a.k.a. the "Green Berets") and has spent his career in a highly decentralized cultural counter-conventionalism, where innovation, adaptation, and creativity have far greater value when compared to the conventional military forces. Grant also graduated from the SAMS program. Chris retired after nearly three decades of military service and continues to challenge the military through military professional education and academic publications. The three of us have practiced implementing traditional military science and found the accepted body of knowledge to be far too limiting, frustrating, and irrelevant in dealing with complex operations. Thus, through critical reflection and breaking out of the traditional military systems-analytic frames, we have attempted to promote and practice military design in novel, distinct, and often deviant (irregular) ways. As our chapter will demonstrate, our efforts have met with somewhat disappointing results, and many of our challenges for encouraging unorthodox design reveal a culturally entrenched institution. We see our adventures with game-changing design as a continuous voyage; yet, we still have hope that our work here will inspire other explorers to take off into uncharted directions.

For the past decade, the U.S. military has been involved with establishing improved management processes and systems believed to enable faster organizational adaptation. However, it has not done well to break out of the old mindset, namely the *systems-analytic paradigm*, which dominates the way the modern military institution (encompassing Army, Navy, Air Force, as well as joint service organizations) *frames* its call for action in military interventions. Instead of applying the same frames of mind and interpreting all future interventions with the same systems-analytic paradigm, we intend to explain and demonstrate how being mindful of alternative worldviews and employing multiple frames can refresh our sense-makings.

Furthermore, the U.S. military is an institution steeped in tradition, rituals, and an insular view of its own history; yet, it can hardly afford to make sense of tomorrow's conflicts by recycling the frames derived from its past indoctrinations (Builder, 1989; Porter, 2009, p. 75).

There are many positions on all of the recent conflicts such as Vietnam, Somalia, Bosnia, Iraq, Afghanistan, or the rather nebulous 'War on Terror.' Our intent for this chapter is to argue these wars are poor sources of analogous framing for the future. We also reason alternatively that the irregularity of modern intervention forbids the otherwise desirable preference for 'decisiveness' particularly encouraged by the archetype of the Allies' success in World War II. There, well-understood enemies were fought and defeated, there was a clear sense of unifying purpose throughout the involvement, and terminal policy objectives were definitively met despite the horrific costs. Figure 10.1 offers a selection of quotes on multiple post-WWII conflicts where the systematic planning archetype failed to produce the expected results. Essentially, we plan to fail without even realizing the fallacy of such rationalizations. Why?

Although the wide assortment of conflict sense-makings presented in Figure 10.1 were quite unique in their own continuously shifting contexts, they all seem to share the same single-framed logic derived from the systems-analytic paradigm. We hope to explain not only how this model of conflict hinders any hope of game-changing design, but that the pressure to conform to the dominant frame serves to socially marginalize or silence any dissenting views or approaches. Our thesis is that there are more ways

---

"The great tragedy of the **Vietnam War** was all of this military effort, great bravery and sacrifice and everything else, was totally unfocused because of the lack of a goal. And because it was unfocused, it failed to achieve the objectives of US foreign policy."
–Colonel Harry G. Summers (U.S. Army), *UC Berkeley Interview,* 1996

"It seems naive now to think that our leadership could have believed that the UN was capable of handling something like **this Somalia mission** and step up and handle a peace enforcement mission."
–General Thomas Montgomery (U.S. Army), *PBS Frontline Interview,* 2001

"If a country can be psychologically damaged, **Afghanistan** was psychologically damaged…My biggest criticism of all of us is that we didn't make a great effort to understand that….We didn't really say, "This is a badly abused nation, and helping this get on its feet is going to be a long-term, difficult, expensive project."
–General Stanley McChrystal (US Army), *The New York Times,* 2013

"Underlying the failed **Basra [Iraq] strategy**, too, was a flawed British assumption that they were good at counter-insurgency. We understand the natives, the generals would tell you—unlike those brutal, clumsy Americans."
–Andrew Gilligan, *The Telegraph,* 2013

Figure 10.1 In hindsight, modern conflicts resist our best intentions.

to frame situations than the military institution's preferred systems-analytic paradigm would normally allow, requiring the ideals of *frame-changing design*. Our argument requires the military observer to interpret events 'in the moment' instead of assuming the role of the cold, calculated 'scientist' who lives by the assumption of objectivity and over-values prediction. Our proposed multifarious way of framing situations requires an appreciation for uncertainty and the recognition that new situations require the invention of new knowledge.

We believe our investigation will help members of other institutions, be they commercial, nonprofit, or governmental in nature, become more aware of similar habits of mind they face when seeking a design approach. For those merely frustrated with institutionally accepted practices, the notion of frame-changing design offers a novel way to untangle oneself from the cognitive 'web of habits' and gain new perspectives. We break the remainder of our chapter into four sections followed by a short conclusion. Consider this outline as something of a conceptual '*plat du jour*' where we introduce frame considerations as we would courses in a French restaurant. Our first section addresses the *institutional paradigm* and provides a short history of military 'systems-analytic thinking' and alternatives that arose in the recent generation of military conflicts. Since the post-Vietnam period of the 1980s, some institutional members of our armed services began to think critically about how the traditional military mindset had been failing, yet our systems-oriented institution continues to seek improvements in the old 'tools of the trade' instead of looking elsewhere. Our second section discusses how interventions require what we call *frame-changing design* as we explore how nontraditional frames of reference may be helpful. Our third section describes current obstacles to frame-changing design in order to give practitioners a better appreciation for issues that will have to be resolved or overcome in the effort to be more effective within the context of both the conduct and avoidance of war.

Continuing with our *plat du jour* metaphor we close our four-course meal with something sweet. Our fourth section presents our *cautious vision of hope* prospecting on how the next generation of military leaders, thinkers, and innovators may transform the institution and expand deeply into a more eclectic, frame-changing design philosophy that we advocate. Our *conclusion* closes this chapter much as we opened it. The military will face more intractable situations and its practitioners' search for meaning should be multifarious—overlapping among governmental, business, and various academic disciplines. Our world is in a constant state of flux and transformation; the way we make sense of it has to keep up.

THE INSTITUTIONAL PARADIGM

How the U.S. Military Prefers to See Nearly Everything

> *It is not true the Pentagon has no strategy. It has a strategy, and once you under-*
> *stand what that strategy is, everything the Pentagon does makes sense. The strategy*
> *is, don't interrupt the money flow . . . add to it.*
> —Col. John Boyd, USAF

> *Actors in war are not billiard balls that only operate according to external pressures.*
> —Patrick Porter, Military Orientalism

The popular view of the military general in Hollywood often seems to be George C. Scott's cigar-chomping, narrow-minded character in the movie *Dr. Strangelove.* Unfortunately, his character, General "Buck" Turgidson, reflected the military institutional mindset more accurately than our flag officers today would give credit. This is not due to some nefarious and conscious effort to drive blindly about the business of executing military operations and strategy, but more of a century's trend emerging from the industrial age spawning of managerial engineering (Romjue, 1997, pp. 10–11, 50–51). Noted military science historian Antoine J. Bousquet summarizes this effectively: As the Enlightenment and Scientific Revolution took hold, reason and scientific method were recruited for the study and organization of all fields of natural phenomenon and human activity, including a quest for the discovery of the fundamental laws governing warfare (2009, p. 56).

Bousquet might have better used the term 'paradigm' to describe this propensity.

By 'paradigm,' we adopt sociologist George Ritzer's meaning: "a fundamental image of the subject matter within a science" . . . that "serves to define what should be studied, what questions should be asked, how they should be asked, and what rules should be followed in interpreting the answers obtained" (1975, p. 7). Anthony Giddens (1994), another sociologist, speaks to the offspring of paradigms, called 'frames,' described as "clusters of rules which help constitute and regulate activities, defining them as activities of a certain sort and as subject to a given range of sanctions." More simply, frames confront the Marvin Gaye question, "What's goin' on?" The term 'paradigm' is an essential concept that becomes a cornerstone to how we need to 'think about our thinking'—how the United States and other Western-style militaries frame situations. Without reflecting on one's own paradigm, these institutions continue to play within the confines of a game that the enemy refuses to play. Henceforth, we define the military's *systems-analytic paradigm* as: the institutionalization of modern military science, [where] one detects mathematical goals of precision that facilitate staff work distribution, categories of resource allocation, and other attractive social architectures that provide the comforting illusion of predictability, control, synchronicity, and other values of scientism (Paparone, 2013, p. 18).

The preferred frame of this paradigm is to view each situation in a systematic way, reducing the confusing mélange of events into understandable chunks, as one would see in an orderly process diagram. By systematic, we expose the belief that complex situations can be broken down into manageable parts, analyzed further, and, through good planning, actions will coalesce toward a unified, desirable end. All this is governed by the weak assumption that if you understand the parts, you understand the whole (Laszlo, 1996, p.16).

A case in point is the U.S. Army War College's publication, *How the Army Runs* ("HTAR" for short). In the HTAR, the Army lays out the analytic systems and control processes that govern how the Army manages its composition—optimizing what weapons it will buy, how it will organize, and how it will fight. Figure 10.2 is an artifact of this paradigm, signifying the extreme in quantitative, systems-analytic thinking involved in Defense decision making. The HTAR is just one example, among hundreds of others in the U.S. military, of this dominant perspective which rationalizes the institution's call for action.

Tens of thousands of civil servants and uniformed staff members participate in these elaborate processes as part of their livelihood and have become so habituated in their thinking they cannot fathom how much the systems-analytic frame of mind blinds them to other possibilities.

There is a cultural history that helps explain how this institutionalization of the paradigm occurred. Much of the systems analysis foundation, found to be successful in World War II, was brought forward into the Cold War by veterans such as Defense Secretary Robert McNamara and his "whiz kids" and continued by others, to include the likes of Secretary Donald Rumsfeld. From aggregating body counts in Vietnam in the 60s to establishing 'measures of effectiveness' about the 'progress' of Afghan security forces in the present, such quantitative measures are not just considered reflections of foreign policy's strategic success, they *become* the strategy (Conklin, 2008, p.4; Naveh, 2004, p. 220). Figure 10.3 provides the parallel relationship between the systems-analytic modeling that the paradigm espouses, and how the military has developed an entire institution dedicated to promoting this scientific-like scheme of the world through centralized decision-making, specialization, systems-designed organization, and career progression through bureaucratic pathways. Making it to the rank to general officer represents the same lengthy and unidirectional path of specialization and progression that the systems-analytic model promotes, spawning a perpetual web of interrelated 'systems within systems' and the stratifications of concepts and correspondent organizational hierarchies that empower them.

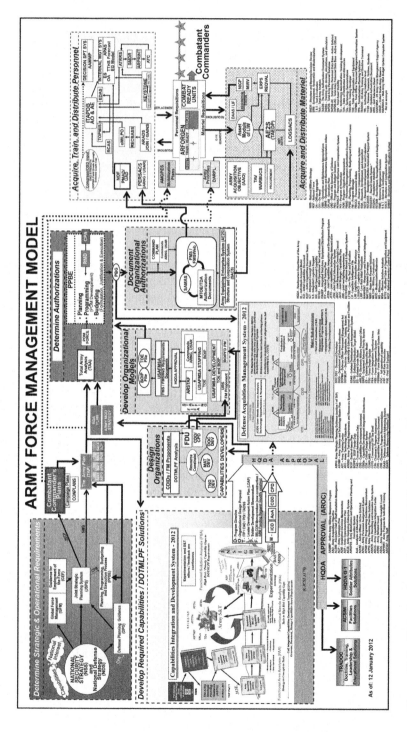

Figure 10.2 How the Army Runs (HTAR). U.S. Army Training and Doctrine Command (2011). *Army force management model.*

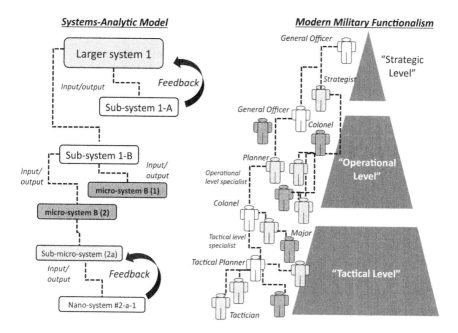

Figure 10.3 Functionalism and the Systems-Analytic Model.

We are not positing that the systems-analytic paradigm is all bad. For example, there is measureable utility that frames results of the Defense Department's budget requirements in incredible detail—what congressional armed services appropriations committees prefer. Such an analytic approach also nests, in theory at least, the President's and his military secretaries' priorities with military plans and operations. The bad part, however, is that the systems-analytic model aligns the institutional mindset to view all situations, no matter how chaotic and disorderly, as if they *were* systematic and reducible to simpler, analyzable components.

In nearly all military doctrine, sequential and reductionist methods inherent to planning frame military action. Military planning, in the U.S. Army called 'the military decision making process' or MDMP (ADRP 5-0, 2012, pp. 2–12) or similar processes used by others services, uses a sophisticated step-by-step systems analysis model: The framework for thoroughly and *systematically* analyzing and understanding any potential OE [operational environment] and all the challenges and opportunities inherent in it consists of the eight variables: political, military, economic, social, information infrastructure, physical environment, and time (PMESII-PT) (Training and Doctrine Command G2, 2012, p. 54, emphasis added).

Similarly, the Army uses the analytic acronym 'ASCOPE' to "categorize the man-made infrastructure, civilian institutions, attitudes, and activities of the civilian population and their leaders" (Center for Army Lessons Learned Handbook 10-41, 2010, Figure 10.4). Joint military doctrine (concepts that apply to the Army, Navy, Air Force and Marines operating together) is also chock full of systems-analytic framing, such as this expression indicates: A system is a functionally, physically, and/or behaviorally related group of regularly interacting or interdependent elements forming a unified whole. One way to think of the operational environment is as a set of complex and constantly interacting political, military, economic, social, information, and infrastructure (PMESII), and other systems.... The nature and interaction of these systems will affect how the commander plans, organizes for, and conducts joint operations. The JFC's [joint force commander's] inter-organizational partners routinely focus on systems other than military (political, economic, etc.), so the JFC and staff must understand these systems and how military operations affect them. Equally important is understanding how elements in other PMESII systems can help or hinder the JFC's mission .... Understanding PMESII systems, their interaction with each other, and how system relationships will change over time will increase the JFC's knowledge of how actions within a system can affect other system components (Joint Publication 3-0, 2011, pp. IV–4).

This systems-framing, which presumably leads to solving 'the military problem', rarely addresses how to deal with a crafty, adaptive, supranational, insurgent network, whether it is on the ground, at sea, or in cyberspace. Such framing is reduced to treating 'variables' in those networks as isolatable, affecting them each with a precise intervention while assuming that all such treatments will add up to victory (Jason, 2001, p. 337; Ahl & Allen,

- SET: military engagement, security cooperation, and deterrence activities
- OPERATIONS: support other governmental organizations (Department of State, US Agency for International Development, etc.) and cooperate with international governmental organizations (e.g., UN, NATO)
  – TASKS: administer foreign military sales program, provide military advice to ambassadors, etc.

- SET: crisis response and limited contingency operations
- OPERATIONS: chemical, biological, radiological, and nuclear consequence management, foreign internal defense, counter-drug operations, combating terrorism, counterinsurgency, etc.
  – TASKS: enforce sanctions, enforce no-fly zones, etc.

- SET: major operations and campaigns
- OPERATIONS: offensive, defensive, stability, etc.
  – TASKS: attack to destroy enemy targets, defend in sector, etc.

Figure 10.4 Range of military operations taxonomy.

1996, p. 1). We have seen over the last dozen years that this reductionist approach to framing situations is not working—the networks appear to be more complex, pervasive, ubiquitous, and chaotic, which defy these systematic, linear thinking approaches to understanding them.

Another example of systems-analytic framing has led to the construction of a range of military operations taxonomy—a doctrinal language that predefines action and neatly establishes clear linkages to training, budgets, and organizational processes and procurement systems (Figure 10.4). To outsiders, this is an intimidating and often highly confusing lexicon that the military establishment uses. It is riddled with many technical terms, acronyms, and specialized phrases that essentially give the appearance to laypeople of a 'hard science' approach to conducting military operations (Zweibelson, 2013). The important message here is that this taxonomy, consisting of mission sets, operations, and tasks, "pre-frames" what action is required and what forces are needed.

The ambiguous world of irregular military involvements is not cooperating with this attempt to regularize frames of reference about them. The systematic approach to neatly breaking apart pieces of a complex situation and labeling them in isolated piles is not helping us win what has become a dynamic, global-, trans-national insurgency. A fundamental question remains: "How can practitioners, finding themselves mired in a 'regularized' systems-analytic mindset, take a critical view of that way of framing as they always seem to face 'irregular' situations?" We posit, as have others (Burrell & Morgan, 1979; Hatch, 1997; Lewis & Grimes, 1999; Morgan, 2006), that a frame-changing approach (1) permits criticism of the single view; and (2) affords alternative views. Frame-changing allows one to first appreciate one's own view of the world, contemplate the intrinsic limitations of that view, see things as others (from outside the institution) might see them, and develop creative approaches to taking action—often through an appreciation gained through multiple frames of reference.

Consider traditional military planning. Instead of following the systems-analytic methods of designing what action is to be taken (i.e., a plan), a more eclectic designer would accept the situation as novel and that they must *learn-in-action* (Lichtenstein, 2000, pp. 48–49).

As part of this alternative approach, the eclectic designer may spend some time with other design practitioners, thinking about what would theoretically constitute 'knowledge' under these circumstances. They could investigate what they know and *how* they know, while being open to *action learning*. We adapt the following definition of action learning—the individual or group endeavor to continuously be engaged in frame-changing "experiments" when engaging in complex situations, all-the-while being open to new and often simultaneously conflicting ways of framing, fully expecting the socially-interactive nature of the situation to produce a state of continuous novelty (Gero & Kannengiesser, 2008, pp. 4–5). Such acceptance of novelty requires learning that is oriented in the here-and-now and actions taken must rely on 'designing-meaning-while-operating.' Today, few countercultural arguments would generate heated, vigorous objection, political infighting, and 'wagon-circling' than the ongoing debate about whether military science should be a rather stable form of knowledge constructed around standardization and systematic models of understanding, or be much more malleable and tailored to the situation at hand (i.e., our central idea of frame-changing design). The prospects for adopting a frame-changing design mentality inside a military systems-analytic culture seem rather dismal (Grome, Crandall, Rasmussen, & Wolters, 2012); although we think there is some hope for the future, as we describe later in our fourth section.

To summarize, the U.S. military institution today should confront irregular situations by considering irregular frames of reference; yet, the prevailing military science is dominated by a habitual mindset that *regularizes* frames of reference. Virtually every U.S. involvement since the end of World War II may be characterized as being plagued with confusion, changing goals, adaptive rivals, and failures nested alongside successes, we must now critically examine the way 'we think about thinking about conflict.' The next question we must address is how thinking differently—with that which we call *frame-changing design*—can assist military practitioners who can, at least temporarily, seek to remove themselves intellectually from the traditional ways of facing novel military situations.

FRAME-CHANGING DESIGN

> *Because even among contrarians, I'm a contrarian. But all of this is just words of bronze, third place rhetoric. What do I really mean when I say we want to shock society into awareness? Do we mean we want more originality and individuality? Less TV, more reading, writing, actual thinking? Less sheep, more shepherd pie? Yes, yes, and a little more pie, please. Oh, and some more sweet tea, too.*
>
> —Jarod Kintz

> *I sent the club a wire stating, "Please accept my resignation. I don't want to belong to any club that will accept people like me as a member."*
>
> —Groucho Marx

When it comes to movies that challenge the imagination and go beyond simply dazzling us with special effects and popular actors, the science-fiction genre is unique in how it invites audiences to imagine well beyond what is accepted as reality. Consider the movie trilogy, *The Matrix*, that was based upon the theory of 'the social construction of reality' (e.g., Berger & Luckmann, 1967; Baudrillard, 1995) known mostly to the few Illuminati of academia. The story wove the theory into a slick action storyline that made for great entertainment. The main character, Neo (played by Keanu Reeves), first realizes through interaction with the character Morpheus (played by Laurence Fishburne), that the world he had 'experienced' throughout his life was an illusion. Morpheus offered Neo the opportunity to break free of his comfortable yet illusive life. In choosing to free himself, Neo realized his former world was constructed by a computer program called 'the Matrix.' Neo journeys back to the fake reality, knowing these visits are but an illusion he can manipulate to his advantage. He discovers his name was an anagram for the 'One' and sets off in a prophetic battle to free millions of others trapped in the Matrix. Although the audience experiences confusion earlier in the movie with Neo as he grapples with whether the world is real or false, at the end of the movie the audience accepts with Neo that everything inside the Matrix world was completely false and could be defeated. Empathetically, the viewers have surrendered their original paradigm (life projected inside the Matrix) and have adapted the new one (life beyond the Matrix).

Of course, we over-simplify this complex movie plot, yet the producers, the Wachowski Brothers, also over-simplified the complex perspectives of the likes of sociologist Pierre Bourdieu who wrote extensively on the need for practitioners to be *reflexive*—constantly aware of knowledge of self and one's position in the social world they realize (Bourdieu & Wacquant, 1992, p. 38) as well as Jean Baudrillard's work to which the Wachowski brothers were most drawn (Baudrillard, 2001). Baudrillard took the idea of 'socially constructed reality' to an extreme position where societies would reproduce their worldviews with no originality. A synthetic computer world, like the Matrix, would have nothing to do with the real world, while everyone in it accepted that it was the only reality.

In that regard, Neo and aspiring military designers such as we three military 'Neo-phytes' have many things in common to include facing complex socially interactive, conflict-ridden environments. Neo's story also parallels the military iconoclast who carries out the work of being an institutional heretic intent on showing others new ways of seeing and interpreting unfolding events. In *The Matrix*, humans were enslaved to a single interpretation of the world, artificially pushed into their psyche that suppressed them from ever discovering that they were not in a real world, but strapped into bio-mechanical life pods and providing power for artificial intelligence which gave humans the illusion of a free thinking- and acting-society.

Figure 10.2—depicting the Army's Force Management model—is a compelling example of how 'Matrix-like' the Army has become and how breaking away from its own 'Matrix' could be refreshing. Just as Neo saw the world outside the Matrix as if he had never used his real eyes before (literally true in the movie), design thinkers that might break free of their own institutional dominant single frame might experience a cognitive equivalent where they 'see' things entirely differently in the very same conflict environment. Within that elegant systems-analytic conglomerate of processes, known as the Army way or HTAR, one can diagnose problems and present solutions through 'gap analyses' with this machine-like frame—determining what has to be done to solve dysfunctions in the system. The HTAR presents a machine-like efficiency that presents a clean, sequential way to consider how the U.S. military designs military interventions under the systems-analytic paradigm: (1) The President releases his national security strategy including the expected missions the military will have to engage in the future; (2) Various departments and component commands engage in their own systems analyses to identify 'capability gaps' to meet the President's objectives; and (3) These agencies determine what solutions (categorized neatly as doctrine, organization, materiel, leadership and education, personnel, facilities, and policy) are needed that they currently do not have to conduct the forecasted strategy and mission sets. The same systematic framing of problems and the analytical framing of solutions also dictate operational planning when emergencies arise in the world. The following case of systematically and analytically determining a materiel solution during the current conflicts is an example of why these frameworks do not work well, no matter how adamantly the architect demands the world behave.

In the last decade, the U.S. Army identified a gap sequentially trickling down from a Presidential policy objective in Iraq and ending up in requiring the standard bolt-action sniper rifle to be replaced by a semi-automatic sniper system. A bolt-action weapon is slower, and requires a manual rechambering of ammunition versus a semi-automatic version that has a higher rate of fire. The principle difference is not that one is 'better' than the other, but that one fires at a faster rate, but also consumes more ammo, has more breakable parts, and thus demanded greater maintenance and care.

In 2007, all units fighting in Afghanistan were notified that they were to turn-in their bolt-action sniper rifles in order to receive the new semi-automatic rifles. The detailed systems analysis seemed right against the reality of the context at hand in Iraq (the dominant 'war frame' of the period) demanded the semi-automatic rifles in urban areas wherein replacement parts and maintenance resources were abundant and the cyclic rate of fire by snipers had to be higher. In the Afghanistan war frame, however, units often found themselves supplied by horseback or air-dropped supplies that came sporadically. The maintenance requirement on the low-tech bolt action rifle was also significantly lower than the new semi-automatic version. The rate of fire of snipers in the more rural areas of Afghanistan was also much lower than in more urban Iraq. When the operational commander in Afghanistan pointed this out to the Army systems analysts, his arguments to retain the bolt-action rifle was rejected on the grounds that the approved gap analysis study concluded that the semi-automatic sniper rifle would be sufficient in all situations. Thus, the process trumped the troops' reality in Afghanistan. The architects of HTAR insisted the game would play out in Afghanistan just as it would anywhere else and any deviation from this scientific 'gap analysis' was merely an anomaly that could be smoothed out with the use of standard deviations and analysis of variance techniques.

This is but one of many anecdotes where the American military institution has grown into a system-analytic culture that makes it very difficult to engage in critical thinking or innovation required to war-frame around the unique realities and morphing, novel situations. Frame-changing practitioners within the military thus face a difficult road to convincing others to accept multiple frames of reference, even if conflicting, to address a variety of contexts. Frame-changers risk being ostracized as institutional heretics. Unlike Neo, we cannot 'bend' physical reality to dodge bullets and win over followers; frame-changing designers instead must apply different approaches beyond the single-frame mode of making sense while acting in the midst of novel military interventions. Grant offers us an anecdote on how the military can at times break free of the system-analytic culture.

I have witnessed countless examples of successful framing-changing approaches. At the typical U.S. battalions (roughly 500 troops) and brigades (several battalions), the right organizational climate can inspire divergent frame-changers to step out and be heard. One such example was a battalion in the 101st Airborne (Air Assault) Division that deployed in 2010 to an area in Afghanistan and attempted to address the most pressing issue they and their Afghan colleagues eventually identified in their sector: a lack of legal justice. When this battalion first arrived they were very concerned to discover that Afghan policemen could not travel in their sector without getting attacked and often murdered. Instead of following a systems analysis playbook and following higher headquarters' disassociated guidance, the battalion commander ordered his staff to seek alternative framings to the novel situation at hand.

The battalion staff, now engaged in an 'action-learning' frame of mind, spent weeks talking to local Afghan tribal elders, policemen, judicial experts, military leaders, charity groups, and others in the area. They quickly discovered that there was no modern judicial system in the area and Afghan policemen were either attempting to intervene into tribal matters or that the existing tribal legal system had been interrupted due to the decades-long fighting that had gone on since the Soviet involvement in the 1980s. There were no jails, no prosecutors, no judges, and the mostly illiterate police had very little training. Many were not even paid regularly, producing a situation ripe for corruption. The people, for the most part, just wanted somewhere to go to mete out justice. The battalion staff, along with leaders from the Afghan security forces came to the battalion commander with a proposal: establish a traveling court that would include representatives of the police, military, Afghan district judicial officials, and U.S. forces who would interact directly with Afghan tribal elders on a very transparent basis in order to address issues fairly within the district where local Afghans felt a sense of injustice. Within six months Afghan police were able to travel freely throughout the district. Insurgent-related attacks dropped to almost zero due to the rejection by the locals of the Taliban narrative depicting the insurgents as the only viable option for justice-related action. The battalion operations officer said it best: Our commander refused to frame the situation initially, instead encouraging us to seek out different viewpoints as to what the problem was. We got different opinions from almost everyone we talked to, but something interesting happened. Because we included all of the various parties during our knowledge gathering, at the end of it all we all collectively had come to a new appreciation: one that no one individual in the group had had before. It was a group learning experience, if you will, and that was important later both because it worked, but also because it got almost immediate buy-in from all parties. (G. Martin, personal communication, August 15, 2010)

The challenge is to somehow incorporate what goes on arguably all across the U.S. military units operating at the local level (addressing the unique context at hand) with the much larger and more systems-analytic oriented departmental level. Of course, with the systems-analytic paradigm, the armed service departments (the Army, Navy, and Air Force) would take the learning that occurred in the local battalion operations area and process it as "lessons learned" (what other communities might call "best practices"). The lesson would become integrated into formal doctrine, gap analysis would reveal a training deficiency in the area of 'jurisprudence operations,' and the practice would be replicated throughout the military through a standardized unit training regimen. Thus, like the sniper rifle scenario, a single and universal solution would be applied to all future problems based upon analysis and a historically directed perspective of learning.

A frame-changing design philosophy rejects such generalizations as fallacies (Simons, 2010); instead, calling for continuous searches for new frames. With new frames would come new complications, deeper questions, and further reflection in an endless cycle of spiraling discussion, learning, and application in recursive interactions. This awareness of change calls for making sense while in action, where tailored solutions in-the-moment cannot be mass produced and reapplied to the next situation. The perspective of learning becomes flipped, where instead of reaching towards the past (Ranciere, 1991, p. xix, 5), the designer continues to orient on the present while incrementally acting.

The U.S. military must not only incorporate the ideals of frame-changing design that enable critical and creative contemplations about far-away operations. Beyond that, the Defense Department should also apply the philosophy to its service departments in how they administer to the forces, encouraging and rewarding critical and creative framing in individuals and smaller groups therein. Such frame-changing there cannot be limited to the preferred systems-analytic paradigm to usher-in innovation. It is, however, one thing to suggest needed change, yet quite another to actualize it. Therefore, we also describe perhaps a more practical and interim solution: actions that can be taken by individuals and subordinate units that, in the absence of widespread institutional reform, can still assist in addressing complex situations. Thus, our next section explores how some frame-changing design has been incorporated in an asymmetrical way as well as how the institution might transform itself to be more divergent in a more overt manner.

DEMISE OF THE MILITARY FRAME-CHANGING DESIGN MOVEMENT?

People don't like to think, if one thinks, one must reach conclusions. Conclusions are not always pleasant.

—Helen Keller

Whoever does not know the truth is looking for it, and there are many encounters to make along the way. The only mistake would be to take our opinions for the truth.

—Jacques Ranciere, *The Ignorant Schoolmaster*

Our imaginative view, described by our concept of frame-changing design, represents a minority voice in today's military services. During the late nineteenth and twentieth centuries, a variety of new paradigms emerged across the sciences and arts. Postmodernism in art and philosophy, critical realism in philosophy, and radical functionalism in sociology emerged, with each spawning the rise of very different views and theories of humans as social beings. These challenged many of the dominant schools of thought—those particularly associated with the systems-analytic paradigm. Yet, while these debates blossomed in academic environments and eventually manifested in the business service industry and other professional practices, the military has been notably resistant to deviating from the powerful grip of quantitative analyses and linear programming methodologies (Ryan, 2009, p. 70; Naveh, Schneider, & Challans, 2009, p. 88). History demonstrates that only after experiencing strategic failure is the military institution ripe for critical reflection and radical change to overcome powerful institutional habits (Cohen & Gooch, 1990). Even then the pressures may not be enough. The American involvement in the Vietnam War can be characterized as doing things that worked in the past, particularly built upon previous victories and universal tenets 'learned' in World War II. Some have argued that the U.S. involvement met with many measureable successes in combat operations yet suffered from strategic failure (e.g., Summers, 1982). After the U.S. pullout, the cherished tenets of military science came under attack by an assortment of counter-arguments, methods, and alternative ways to frame situations. But these were short-lived and the U.S. military culture soon reverted back to the mainstay perspective associated with system-analytic design and devoted attention to high technology warfare (Romjue, 1984; Builder, 1989). The systems-analytic model, geared to a universal design technique and application, demanding uniformity over unique-, tailored-approaches, remained a dominant aspect of the military institutional approach to virtually everything (Weinberg, 1982, p. 12). After more than a decade of post-Cold War conflicts, we may be confronted with the beginnings of what may be construed as post-Westphalian (i.e., a neo-nation-state system), evidenced by failed nation states and transnational extremist movements (Rosenau & Czempiel, 1992). Henceforth, military design began to take serious form in the initial intellectual exploration by the Israeli Defense Force (IDF) in the 1990s particularly in its think tank, the Operational Theory Research Institute (OTRI). Some influential military theoreticians began to discuss paradigm-breaking ideas associated with 'design' (e.g., Naveh, 2004; Wass de Czege, 2010). This initial work from a small number of paradigm-breakers drew heavily from urban policy design concepts that attempted to address complex, intractable social problems (e.g., Rittel & Webber, 1973; Schön & Rein, 1994) and from the emerging science of complexity and complex adaptive systems (Sanders & McCabe, 2003; Osinga, 2007; Bousquet, 2009).

Soon, the U.S. military began to take official notice, and by 2008 introduced preliminary design concepts into experimental concept papers as well as advanced planning schools such as the highly vaunted U.S. Army School for Advanced Military Studies (e.g., U.S. Army Training and Doctrine Command, 2008). As official doctrine began to convey design concepts to the larger military audience, the doctrine writers, who had vested years of work that was ingrained in them by the systems-analytic paradigm, unfortunately changed the originating eclectic underpinnings of new design ideas to create yet another systems-analytic approach to design. Thus, through this reinforcement of the dominant frame, strategists and operational commanders also contributed to the countercultural 'threat' of frame-changing design by corrupting the original ideas and subordinating them to the comfort of the systems-analytic rubric. The doctrine writers eventually removed all reference to other paradigms and opposing theories that would have been intellectually revolutionary for the community of practitioners. Now 'design' is taught and trained as an applied 'step by step' analytic approach—the U.S. Army calls 'design methodology.'

By 2009, official design doctrine was imbedded into existing systems-analytic processes and taught as just another planning tool in the professional military education system. 'Doing design' arguably became an 'add-on' to existing sequential decision processes. As this military school of design expanded, U.S. combat training centers as well as specialized high-level staff teams adapted design for the majority of major training events; albeit, in this 'neutered' state where frame-changing design became lost to the dominant systems-analytic paradigm. In 2010, after only about a decade, from a promising start to a culmination of a much-diluted design philosophy, the potentially revolutionary conceptualizations were reduced to just another systems-analytic method. Indeed, some have decried that the frame-breaking form of design is now dead and argue that the military, particularly the U.S. Army, needs to 'get back to the basics' and 'forget about all this philosophical mumbo-jumbo.' We three observers have experienced this so frequently that we hesitate to even use our view of 'design' in military circles for fear of career retribution. Many opponents use the term "design-istas" or "secret decoder ring club" to marginalize and type-cast those who propose the countercultural approach to frame-changing.

At the time of this writing in late 2013, we believe the military's institutional failure to grasp the underlying need for multiple frames drove the near death of a true frame-changing design philosophy. Eclecticism calls for conceptualizing complex situations using a variety of perspectives developed through a diversity of academic disciplines and cross-cultural perspectives to spur collaborative forums for debate. This implies acceptance of competing values that may triangulate new meanings while acting in the mélange of indescribable, ever-morphing, socially-interactive situations. This type of thinking for the traditional military mind challenges cherished values within the military profession, such as "We are competent and know what to do." In short, the systems-analytic paradigm ate the eclectic paradigm for lunch. The only reason that frame-changing design remains a viable concept for the future is that, the systems-analytic perspective has once again demonstrated strategic failure in Iraq, Afghanistan, Libya, Egypt and we expect with current issues with the Syrian civil war and the Russian annexation of Crimea and possibly other parts of Ukraine.

Although our last few paragraphs cast a somewhat jaded perspective on where the military is going institutionally with respect to frame-changing design, there are many reasons to be optimistic on where the military will go in the next decade or so. Failure is indeed the best teacher, as it helps overcome the sanctions that the systems-analytic frame has yielded. Military practitioners are trapped in a Matrix-like false reality and are hopefully beginning to recognize that the 'men in suits' demands for systems analyses are not resulting in the effects expected. There has to be a change in mindset and we are cautiously hopeful. This is the topic of our next section.

A CAUTIOUS VISION OF HOPE

Five percent of the people think; Ten percent of the people think they think; and the other eighty-five percent would rather die than think.
—Thomas A. Edison

This is our most desperate hour. Help me, Obi-Wan Kenobi; you're my only hope.
—Princess Leia, *Star Wars Episode IV*

We believe there are still prospects for our proposed frame-changing design ideas. Signs of hope are few, but encouraging. For example, we note the open discussions on unofficial internet websites such as that offered by the *Small Wars Journal*, and the recent expansion of novel design articles in the mainstream of military academia such as the journals *Prism* and *Military Review*. We also are hopeful that the gradual increase in senior officers who are intimately familiar with novel settings after participating for more than a decade of war. This may lower their expectations about the systems-analytic doctrines, which in turn may open the door for truly dissimilar frames and critical frame reflection. Indeed, we have observed in our own experience with irregular situations, that there is a growing frustration with doing things regularly, 'by the book' and in 'fighting enemies that do not play by the rules.' While the institution continues to insist that we just need to improve on the systems-analytic models we have in place, the frame-changing aspects of design help us question and even dismantle the 'rules' that govern these habituated interpretations.

There are more signs of hope. Like Columbus setting sail and expecting with every fiber in his body that they will open a 'trade route to India,' sometimes the emergence of novel discoveries while 'floating in uncharted waters.' Ben, the lead author of this chapter, offers a tale of designing outside of the doctrinal boundaries in Afghanistan: As an operational level planner in our NATO training unit in Afghanistan, I had the tremendous opportunity in mid-2011 to address a sticky question first posed by senior policy makers in Washington. They asked us what we thought the future threat environment might look like in five years in Afghanistan, what the future Afghan Security Forces ought to be at that point, and how we might adjust and tailor the current force between now and then to deal with this future scenario. From our experiences in the messiness of our efforts in

Afghanistan, this was a seemingly ridiculous "hypothetical question" for us to attempt to answer. Out of sheer happenstance, as one of the junior planners on the team, I simply volunteered to answer the question when our planning group finished complaining about the impossibility of it.

My boss quickly gave me the standard 'left and right' limits on what he wanted, and what deliverables he expected when I was complete. No one at that time realized how this little project would soon blossom into a monstrous planning effort that would warrant attention from the highest levels of strategy, political policy, international cooperation, and high stakes military power plays in the international media. My Colonel, a career strategist for the Army, instructed me to "do the research, build a project team from whoever in the NATO headquarters you need to draw from, do some formal 'war-gaming' using doctrinal (yes, systems-analytic) models, build some briefings, give us updates, and produce a 'white paper' on the findings." This process exemplifies the standard life-cycle of most high-level military research projects and was nothing unusual. However, I knew the question was impossible to address with the established military systems-analytic, problem-solving methods that doctrine demands. On the contrary, much of what that process would have produced would be disastrous, with as much efficacy as a Shaman rain dance. Having written my thesis on 'design theory' at the U.S. Army School of Advanced Military Studies earlier that summer just before arriving to Afghanistan, I was confident that unconventional design approaches I had studied would prove to offer better frameworks.

Prior to leading this planning team, I experienced some rather hostile positions by some curmudgeons on concepts associated with design thinking, particularly from those highly invested in our traditional systems-analytic paradigm—where 'hard' science, data, charts, algorithms, and metrics defined everything of value. The mere discussion of an eclectic approach to design—delving into postmodern philosophy, interpretive theory, and other 'artsy' concepts—resulted in sour faces and closed-minded rejections. The only progress I found was in the use of the 'bait and switch' method of a magician, where after you interest the audience with the card trick, you then show them how you did it. By using many of these unmentionable concepts in "the trick," I found that I could disable many of the institutional frames of reference in place that opposed what frame-changing design offered. I applied that approach here liberally by starting with an extremely small planning group.

Our small team, comprised more or less of fellow eclectic design thinkers, set off to use a fusion of postmodern philosophical concepts from some French thinkers; some scenario planning models drawn from the Exxon Company in the 1970s, semiotic square concepts from an assortment of theories (Corea, 2005), and interpretive organizational theories I had learned from reading the multi-frame approaches from researchers such as Professor Mary Jo Hatch (1997). None of these ideas were present in official planning doctrine and if introduced to outsiders of our little cabal, we knew these would trigger immediate rejection or disinterest. Instead, we pulled a trick—we incorporated these non-doctrinal approaches within the existing systems-analytic frameworks adopted by the institution.

By taking the format of familiar war-gaming structures, stripping out the concepts and replacing them with these design concepts, we were able to fashion a hybrid approach that worked as an intellectual 'Trojan Horse.' Covered by familiar terms and concepts, once we introduced our larger working group team to the initial design concepts, they perceived the outcomes of what they thought were traditional war-gaming methods and other systems-analytic practices. Now our core design team could immediately go into deep discussions on how nearly everything 'hidden under the magician's sleeve' was entirely different and unlike familiar war-gaming, but we needed to get the larger group invested intellectually into the project without triggering institutional resistance. We avoided talking about paradigms, non-linear thinking, swarms, semiotic squares, paradox, systemic versus systematic, interpretive reasoning or Hatch's cultural approach to assumption-value-artifact-symbol relationships (Hatch, 1997). Instead, we framed nearly all of this in generally simple PowerPoint slides that outlined in a familiar analytic methodology how we would proceed. We did introduce semiotic squares, but did so in a way that applied a linear-like process so as to simplify the underlying nonlinear logic.

The deceptive nature of some of this process might alarm some readers who might claim these are unethical practices. Granted we could probably have sat down many of those in our group and had extensive discussions on design, and likely won many of them over to our approach. However, we had less than two weeks to do this entire working group while also doing our regular jobs in a combat zone. There simply was no time, thus we moved past debating design to simply tailoring design concepts for use with non-designers and cast within familiar methodologies to gain their buy-in.

In the end, our larger planning group, chaired by our inner team of design-magicians, produced an accepted proposal that ultimately went to the highest levels of political decision making by the fall of 2011. By early 2012, the concept was embraced by the Coalition, and although it was just a concept for the Afghan Security Force beyond 2015, it subsequently drove future planning processes that likely continue today in a vastly different, advanced form. Our mission as we saw it was not to "solve" the problem for a post-2015 security force, but to get the concepts moving with an initial and highly speculative frame from which our counterparts could improvise. The methods of frame-changing design we used were not in what we represented in that final proposal; rather, we quietly represented the underpinning philosophy. We were successful in disguising some highly conceptual design concepts into systems-analytic based war-gaming, capitalizing on a highly diverse group of participants, without our project team devolving into dysfunctional group dynamics.

As Ben concludes, the game-changing design his team was a part of was not necessarily *what* his team produced, but in *how* they produced it. One possible outcome of this sort of frame-changing would be to redefine how military practitioners define and make sense of the term 'to win.' Under the systems-analytic paradigm to 'win' is to reach a clearly defined and measurable goal that one planned prior to acting. 'Winning wars' has a powerful history and symbolism for our military, whether documents are signed upon a battleship or a sailor passionately kisses a nurse in New York City. 'Losing' is equally symbolic, whether reading of our White House being burned by the British or of desperate Vietnamese partisans attempting to hang onto the helicopter landing skids as our Saigon embassy is evacuated. Yet, in today's flawed understanding of the past, we yearn for a "return to normalcy" wherein every war was a "good war" with clear objectives, ending, and outcome. That constructed reality is like the Matrix—reality was never that "clean" and we should expose those narratives as mythologies that seek to describe the future in terms of the past.

A frame-changing view would consider ways to frame those constructs that include more open-ended, in-process views of winning that might describe an on-going and unforeseeable-ending effort. This view might better match an abstract concept like 'war on terror,' 'war on poverty,' or 'war on drugs.' Still another way could be to drop the term *win* altogether and possibly adapt the language of the major players in the region in which the U.S. military is operating or supporting. For example, the U.S. government could follow its rhetoric and refuse to turn "political" efforts into "military" efforts and thus possibly avoid the need to characterize something as needing a win or even a military objective. Military efforts, due to the military's penchant for a systems approach, requires military objectives and this very systematically focuses all entities of the U.S. government on military action and in turn the need for a military win, even in situations like Afghanistan where even those in command acknowledge (along with everyone else) that a military solution will not work by itself. Afghan ministers that Grant and Ben worked with often referred to most of the Afghan Taliban as their "confused cousins" as opposed to "enemy forces." Frequently, they blamed most acts of violence on foreigners and invaders, viewing the majority of Afghan citizens as peaceful (Arbabzadah, 2011). These language distinctions, coupled with nationalistic, ethnic, and cultural considerations unique to Afghanistan tend to change the meanings and concepts associated with our military's dominant systems-analytical model and our own lexicon. We end up talking past each other despite our own best efforts.

If the United States had used Afghan language to describe the effort then the military probably would not have needed a "win" when all they faced were "confused cousins" as opposed to an enemy. Changing the language could have potentially changed the entire course of the operation and focused all efforts onto what the U.S. government and even military units called for during counterinsurgencies: more of a political versus a military effort. It is scary how much an organization's conventional labels and institutional 'understandings' can focus efforts in unproductive ways. A further example of this frame-changing is described by Grant, another chapter author, in a description of another planning effort in 2010 at the NATO Training Mission Headquarters in Afghanistan (NTM-A): In the early portion of 2010, NTM-A was struggling with how to best train and equip the Afghan Army and police to support the overall U.S. and NATO effort. During that time there were considerable political pressures to entertain the notion of cutting all ties with the Afghan police due to the prevailing belief that the police were doing more damage than good and that the allies could get more if they consolidated their efforts around the more competent Army. However, equipped with the history-based doctrine of counterinsurgencies, NTM-A senior officers were convinced that police forces were more important than military forces in combating insurgencies, therefore NTM-A promised U.S. and other NATO-member political leadership that they could show progress with the Afghan police. When asked for a time estimate to demonstrate progress, the politicians answer was "90 days."

After that meeting, NTM-A leadership went back to the general staff and asked how we would show progress with the Afghan police in 90 days. The staff quickly decided that of the various divisions within the police, the Afghan Civil Order Police (ANCOP), since they were the smallest element, would be the best unit to concentrate on. One consideration was that the ANCOP was only about 5,000 members on paper, and more like 3,000 in reality. Secondly, the ANCOP suffered from tremendous attrition—as high as 76% in some months. It was felt that fixing ANCOP in 90 days—at least offered the political leadership some positive metrics that showed improvements.

At the same time, the rest of the command was involved with brainstorming how to "fix" the ANCOP; our informal design group within the command was comprised of members of all staff specialties and the majority of the nations involved in NTM-A took on the same problem independently as a check on the rest on the combined command results. The problem, however, was that the command had a very narrow path for decision-making, vested with a select few 'powerbrokers' within the command. We realized that in order to make our concerns heard, even though we were scheduled to have an audience with the commander himself, we would need to gain consensus about our conclusions through the various powerful staff sections prior to our meeting. In essence, despite the assumed autocratic nature of military command, we would have to 'play power politics' to get the various staff sections in agreement on our understanding of the situation and ideas for going forward.

Thankfully, our design group had been consciously formed out of members of the various staff sections, and so as we met and discussed the AN-COP issue, these members periodically went back to their powerful staff section officers and interjected our thoughts into their deliberations. Eventually, these conclusions filtered out throughout the command and were mostly bought-into long before we met with the commander; in fact, his comment at our meeting was, "I've actually heard many of these recommendations already from the rest of the staff," which was taken as a snide rebuke by some of his inner staff and may have been meant as one by him; but we took it as positive feedback that our political strategy was working: the ideas were more important than the credit. So, framing our project as a political activity was important.

How we went about getting our frame-changing ideas accepted, however, was only part of the design effort. The other half was framing the issue at hand in the crafting of recommendations based in an eclectic approach that forced us outside our institutional mindsets. For this we reached out to as wide and diverse an audience as we could muster. We sought out the audience of Non-Governmental Organizations like Afghan and foreign human rights groups. We met with as many Afghan policemen and Ministry of Interior officials as we could. We talked to Afghan interpreters, Afghan local workers, Afghan businessmen, others within the Afghan government, Italian and German embassy officials (the Italians and Germans were involved in the judicial system effort in Afghanistan), and officials from the European Union Police Mission (EUPOL). We also toured several Afghan police installations and training sites and talked to trainers and students as well as Coalition trainers.

From all of this 'action learning' we added a very key component: to investigate why the political actors asked the question in the first place and include trying to understand the interaction of political leaders with NTM-A command styles linked to in the task to "show improvement with the Afghan police in 90 days." Before we could even begin to understand how our recommendations needed to fit into the overall politicized situation, we had to understand why we were really in Afghanistan at the moment. To do that, we had to go much deeper than reading the publicly released strategy and official policy statements from our leaders. We had to look at what we knew about the politics of our situation—both domestic pressures in the United States, within Afghanistan, among every European nation involved in the effort. After much discussion and questioning and a lot of assumption-making, we concluded that the reason we were in Afghanistan was that the respective NATO countries' populations could not articulate a reason to their populations why they should be fighting in Afghanistan and at the same time they also perceived international political costs to pull out unilaterally. This complex frame of reference not only helped us make more sense of the situation at hand, but also helped avoid crafting recommendations that we knew would not get past the first level of approvals.

We also discovered that the ways the Coalition was measuring success was bumping directly into the reality of the Afghan domestic power struggles that needed to play out one way or another so that stability could be established. Instead of facilitating solutions that would have encouraged stability, our way of measuring success forced us to do the opposite—arrest the development of stability mechanisms. In short, our culture and political system were unwilling to accept the pain and suffering necessary in the interim in order to get to some stability, and thus we were fighting a status-quo holding action: realizing that a step forward resulted in one or two steps back. Our recommendations were two-fold: for the long-term we needed to synchronize the development of the Afghan judicial system with the training and deployment of the Afghan police as well as take a more hands-off approach with the police in order for them to go through the pains of learning and growing on their own (as opposed to "parenting" them—note the frame). This would naturally cause some pain in the short-term, but we felt that we could mitigate that with the ANCOP by giving them a "red-cycle", or down-time in between their long-term deployments. At that time the ANCOP units were patrolling and fighting mostly non-stop and mostly in the South, which was problematic for a force mostly composed of Tajiks from the North and a force that suffered disproportionately than the rest of the police. Instituting a "red-amber-green" rotational cycle would also help in that during the "red" time they would be off and the "amber" time they would be training up for their "green" deployment phase. This was similar to the cycles U.S. units aimed for. In addition, we proposed paying the ANCOP more (like we did with the Army's commandos), partnering them better, and assisting them in getting home when they were in a non-training cycle.

As most of our short-term recommendations were successfully implemented, we felt that we were at least partially successful in our design effort. What was enlightening to us, however, was what we discovered of our own organizational politics that existed 'beneath the veneer of systems-analytic rationality.' This knowledge helped us as we went forward and tackled other intractable problems: such that we would have been most likely very frustrated if we had not attempted to better understand our own institutional bureaucracy and its competing values. We termed our effort 'insurgent design' because at one point we received word to not mention the word 'design' around the commander since he was not a proponent of the new associated ideas. We did learn that 'insurgent design' was an appropriate name, as we were able to get our recommendations to "bubble up" through the command in such a way that the commander did not know we used frame-changing design nor that we had come up with most of the recommendations before 'leaking' them to the rest of the staff. We 'infiltrated' our ideas 'covertly' throughout the command and no one seemed to suspect that they were ours when the 'irregular' project was turned in.

The fact that neither of our stories thus presented focus on decisive action or "winning" is something that we see as key to design efforts. Frame-changing design is a social process; hence, it is never 'over' and thus assigning arbitrary 'end-states' and 'objectives' and 'backwards planning' in a linear fashion is not conducive to a multi-framing approach. In both cases a frame-changing approach allowed the design teams to see the futility of setting one "victory" condition instead of attempting to see how different groups would see things. Instead, designers aimed to ensure we incorporated means with which to continuously re-frame as we noticed feedback that did not make sense with our original frames. One way in which design may be incorporated into the military much better than in a centralized fashion will be for individuals to be exposed to the concepts and then take them on independently in their own ways. An inspirational speech described in our final section by our third author, Chris, has not only possibly inspired many graduating officers from the Army's Operations Research Systems Analysis (ORSA) course, but has inspired some members of the U.S. Army Special Forces and thus potentially many members of foreign militaries they advise in the future.

ORSA is a field of intense study for a specialized group of military officers. ORSA theories and methods epitomize the systems-analytic paradigm that the military institution has adopted as its foundational science. As part of the evolution of 'McNamarianism' we described earlier, ORSA frameworks for military design rely on operationalizing variables, measuring independent variables effects on dependent variables, conducting sophisticated quantitative analysis, creating quantitative decision support tools, and finding 'optimization' for the use of scarce resources. If frame-changing design represents a direction where military practitioners are encouraged to reflect critically and challenge many of the institutionalisms of analytical-based modern military science, we would offer that the field of ORSA represents a single paradigm; hence, sits in opposition to the ideas we have presented in this chapter. While we would endorse the use of methods drawn from the systems-analytic paradigm as part of a multi-frame endeavor, we believe the ORSA-dominated science would reject our frame-changing, eclectic ideas.

As a fan of frame-changing design, Chris found himself in a situation similar to Ben's story—where design might operate as a 'Trojan Horse'—and as a corollary to Grant's characterization of 'insurgent design.' Each of us has attempted to breach the institutional defenses on that otherwise close discussion. Chris had the unique opportunity to insert many of our chapter's themes into the ORSA class graduation address, in yet another prime example of bending the debate towards reflective practice and 'thinking about how we think.' We share excerpts of his address below, to highlight these reoccurring messages: What I admire most about the Military ORSA profession is that it is arguably the only field that employs sound scholarship and research methods in support of both the operational fight and force management. You have heard the drum beat from many guest speakers who have spoken to your class—that your work translates into theater-level and strategically important decisions. That responsibility is nothing short of awesome.

As you have experienced we have tried to weave more critical thinking into the ORSA qualification course than in previous versions of the course. I believe this is prompted by the idea that you will no doubt run across research projects that are so novel that the tools you have been taught and have used in the past may not apply... or that you'll have to make adjustments to them and use them in ways not previously considered.

Chris goes on to introduce the intellectual 'Trojan Horse' of design to the ORSA graduates, mindful of their preferred systems-analytic paradigm: My theme is about relationalism and relational thinking. Relationalism is a philosophical concept that recognizes the study of human relations as too complex to reduce to discrete variables and linear causality. Human relations are replete with dynamic interrelationships, paradoxical values, and ambiguity. Studies of human relations are necessarily subject to multiple interpretations or perspectives.

I can immediately sense the consternation from the audience—this thought system is obviously quite different from the traditional ORSA approach that calls for operational definitions, root causal analysis, and simplification. You must be asking, who is this odd man at the podium who is calling for just the opposite? Please bear with me. . . .

Chris goes on to frame some examples of frame-changing thinking: I was pleased to learn in a recent social gathering of ORSA officers that there are some who are delving into new ways of framing and approaching complex issues, such as through social network analysis, swarm theory as applied to the use of remotely piloted vehicles, and a movement toward qualitative or interpretive forms of research. This movement is key, in my judgment, to making new sense of complex and seemingly chaotic aspects of our social world that includes military interventions.

As we intended to propose in this chapter, Chris attempted to encourage some critical reflection on how to make sense of the world, and how we might reflect upon things we take for granted (as facts). The ORSA student represents the pinnacle in 'fact-checking' for the military, as their research and quantitative outlook tend to assume a mantle of infallibility in today's military institution. Chris used his speech to address perhaps the fundamental cornerstone of the ORSA mindset: What constitutes factuality? Thus, he engages the class not with a methodological discussion on what data to use or what quantitative process to apply, but with an ontological debate challenging what the ORSA graduate considers 'real' versus 'not real: At one end of the spectrum we have natural facts. These are pretty easy to understand. There is little ambiguity, for example, when we see a mountain. While we might disagree between languages and cultures what to name it and what it might represent, it is still "there" in an empirical way. Things get a little murkier when we move toward social facts...

A spear is a type of social fact. It is "there" but any member of a secluded culture who might see it for the first time might not have the same sense of "natural factualness" about it as would a native who made it and uses it. Yet, an outsider would probably be able to intuitively figure out for what and how it is used. The reality of the spear is arguably less socially specific enough as to be interpretable even though it exists as a physical reality. Its functionality is only implicit to the society that created it; hence, the name artificial facts or "artifact."

At the far end of the fact spectrum are another sort of social fact—these pertain to socially agreed to concepts. Without that social agreement, they would simply not exist and certainly not be meaningful to any outsider. My plastic card [Chris shows his credit card] is a good example of this type of fact. It is not implicit to all cultures (say in the remote parts of the Amazon River) as to its purpose. Even if familiar in our culture, there are many [of these sorts of social facts [one would] need to understand that this is not just a piece of plastic. This card represents a complex web of meanings—the concept of digitized money, virtual banks, debit cards vice credit cards, PINs, activation, stolen card reporting, credit insurance, cancellations, and so on.

With access to the global internet, culturally specific facts are disappearing. But this should not dissuade you as an ORSA to critically recognize and understand that [these] facts are still complex human inventions where meaning is derived from a complex network of socially constructed meanings. Operational definitions become problematic.

Chris is playing the role of the 'insurgent designer,' encouraging critical reflection for what is typically a group of military thinkers that are focused on the facts, unaware of the construction of the social reality in which they are living. He continues with the discussion, moving critical reflection towards creative application: As ORSA professionals, I believe it is imperative for you to delve into social construction theory. We all heard the General say earlier this week that ORSA is all about presenting the facts. I challenge you to be able to reply confidently, "Yes Sir, but what kind of facts?"

My next topic... includes some notes about categorical thinking. I think the point I make here is important for you as a well-rounded researcher.

Caution, these ideas are contrary to the *operationalism* you have been taught. As we can see in the emergent postmodern world, categorical boundaries of our institutions are becoming more and more ambiguous. What is a nation state? What is marriage? Is the national treasury an illusion or reality? What constitutes law enforcement with respect to counterterrorism? What is war?

The postmodern condition (in this case, evidence of a post-institutional age) can be unnerving to the more conservative-minded. Definitions, rules, and correspondent values that constituted what we thought were solid institutions are unraveling. It's hard to make sense where we thought we had stable sensibility. What I propose to you today is that part of our sense-making problem is that we are tied to categorical thinking.

Chris then covers many of the points we make in this chapter, where the systems-analytic paradigm spawns the usual quantitative framing approach that tends to limit, and in some cases prevent, understanding the emergence of complex, messy conflict situations as they unfold. While not rejecting the systems-analytic mode, he remains open to other ways of framing: Frame-changing can paint a more realistic picture of the paradoxes of living in a complex society. As a society we want security as well as liberty (and we slide along that scale frequently). We want the efficiencies of capitalism and we simultaneously care about equity. People and organizations are both internally and externally focused. People and organizations are highly adaptive while at the same time well-controlled. Accepting paradox is not an either-or proposition. The study of paradox presents a relational way to appreciate the yin and yang of our social life.

What if we were to envision our future organizations along continua rather than categories? Perhaps we could better respond with a range of complex organizations that works in response to the chaos and complexity at hand. Sometimes you need high adaptation and improvisation as in response to a surprise catastrophe. Sometimes you need machine like reliability as in running nuclear power plants, aircraft carrier landings, and performing vehicle maintenance.

Chris then provided several books and theories for the audience to research, in case his intellectual Trojan horse triggers any interest and self-reflection in the group. He closed the speech with a summary that is fitting as a preamble to our conclusion: I have burdened you with the ideas of frame-changing. Let me sum up by saying that frame-changing seeks to sense-make with the ideas associated with chaos and complexity, accepts the ambiguous nature of social facts, and instead of categories, employs paradoxical continua as "normal" to the processes of inquiry. To inquire relationally is to critically reflect with multiple points of view, not having to settle on one. Frame-changing acknowledges that social phenomena are not 'homogenizeable.'

CONCLUSION

If you make people think they're thinking, they'll love you; but if you really make them think, they'll hate you.
—Harlan Ellison

I guess you guys aren't ready for that yet. But your kids are gonna love it.
—Marty McFly, *Back to the Future*

We have presented our frame-changing design concepts in the hope that we have convinced the reader that there are more ways to frame situations than our institutions would normally consider. We first described how our institution, the U.S. military, has developed a mindset which powerfully dissuades practitioners from deviating from the norms generated from the systems-analytic paradigm. We offered an alternative paradigm we call frame-changing design and provided real-world examples of how this ideal can be transformed into workable ideas. We introduced the survival strategies of present-day frame-changing designers who like the 'Trojan Horse' or an 'insurgent' found ways to learn-in-action. Our prognosis for survival of the design movement in the U.S. military is pessimistic, at least in the short term; however, there are glimmers of hope that frame-changing design thinkers will emerge from over a decade of experience in complex military interventions. This will be an uphill climb as the institution is clearly infatuated with the science of operations research systems analysis, the artifact of the inculcated systems-analytic paradigm. Those that promote frame-changing designs are apt to suffer some career risks because the military tends to confuse critical reflection with institutional disloyalty.

Despite the dangers ahead, we are cautiously optimistic that frame-changing designers will be an influential part of the next generation of senior military practitioners. Game-changing design in military applications is not about getting better at fighting wars more effectively so that we might win the next one. It is about understanding how and why we tend to fight wars ineffectively, why we often approach a conflict from the wrong direction with faulty expectations, and how we might limit or prevent war from even beginning. Our hope is that as we reflect on the inadequacies of employing the systems-analytic approach to our recent experiences in novel, ever-changing military interventions, the institution (our 'Matrix') will be sufficiently reflexive to 'see' a need for frame-changing design.

Design is risky, often dangerous due to the very notion of 'game-changing.' With the massive size and deeply ritualized institutionalisms of the military establishment, we cannot expect the aircraft carrier to simply turn on a dime. Although it takes years or even decades for technology and tactical methods to cycle out as innovations and discovery creates better replacements, military ideas and concepts are decidedly more rigid (Schön, 1963, p. 83). Our chapter does not argue changing this tactic or that method on targeting an insurgent cell in Somalia. Rather, we took aim at the entire way our military prefers to make sense of the world around it. Instead of a methodological debate, we seek an epistemological one. Unlike the local effects of targeting a single system or concept where a sniper rifle manufacturer might rally against us, we are upsetting the largest apple cart in our institution—the philosophy behind how we make sense of the world.

In no small way, challenging how an institution frames is perhaps the greatest debate one might have; if successful over time, designers will ultimately change every aspect of "the game" of military operations, from the rules to the players to the very notion of winning and losing in the twenty-first century. It would be naïve to think that change could happen rapidly and with minimal disruption. Unfortunately, philosophical discussions alone do not suffice. What does effectively prolong this debate is the perpetual failure of the current system. Although those in defense of the dominant single-frame approach will continue over the years to come to attack design thinking, and cast the blame of failure upon military operators instead of the system itself, we think that this trend in time will reverse. What is particularly disappointing is that failure does not come in a vacuum in military conflicts. Unlike corporate boardrooms where failure comes with either a pink slip or the unfortunate merger with a superior competitor, military failure brings the tragic stink of death, destruction, and the tremendous weight of post-conflict complications, whether they come in financial, physical, or emotional forms. While the current generation of senior military leaders is inevitably tethered to the success or failure of current conflicts, the next generation of military leaders is already forming novel, dissimilar ideas on where to move next, and what to discard.

Game-changing design in business ushers in new ideas, profit, and an exciting new world of options for new adventures. Game-changing design in academic disciplines brings with it the opportunities to increase our understanding, expand our knowledge base, and open previously locked or hidden doors to human expression and awareness. These are all wonderful, exciting things. For game-changing design to work in future military applications, we have to nudge our institution towards reflecting critically and creatively upon failing practices, no matter how self-relevant, traditional, or essential we deem them to be. This is no easy task. The next generation of young officers, predominantly in the Company and young Field Grade range, are the thinkers that have experienced the current military sense-making approaches. These fresh perspectives represent the pending emergence in novel adaptation, improvisation, and radical discovery that will change the game, perhaps to levels where we might not imagine.

REFERENCES

Ahl, V., & Allen, T. F. H. (1996). *Hierarchy theory: A vision, vocabulary, and epistemology.* New York: Columbia University Press.

Arbabzadah, N. (2011, June 30). The Kabul hotel attack was destined to happen. *The Guardian.* Retrieved August 30, 2012, from: http://www.guardian.co.uk/commentisfree/2011/jun/30/kabul-hotel-attack-.

Baudrillard, J. (2001). *Simulacra and simulation.* (S. Glaser, Trans.). The University of Michigan Press.

Berger, P. L., & Luckmann, T. (1967). *The social construction of reality.* New York, NY: Anchor.

Bousquet, A. (2009). *The scientific way of warfare.* New York, NY: Columbia University.

Builder, C. (1989). *The masks of war.* Baltimore, MD: Johns Hopkins University.

Burrell, G., & Morgan, G. (1979). *Sociological paradigms and organizational analysis.* Portsmouth: Heinemann.

Center for Army Lessons Learned (2010). *CALL Handbook 10-41.* Retrieved September 25, 2013, from: http://usacac.army.mil/cac2/call/docs/10-41/ch_1.asp

Cohen, E. A., & Gooch, J. (1990). *Military misfortunes.* New York: The Free Press.

Conklin, J. (2008). Wicked problems and social complexity. *CogNexus Institute.* Retrieved September 15, 2013, from: http://www.cognexus.org

Corea, S. (2005). Refocusing systems analysis of organizations through a semiotic lens. *Systemic Practice and Action Research, 18*(4).

Gero, J., & Kannengiesser, U. (2008). An ontology of Donald Schön's reflection in designing. *Key Centre of Design Computing and Cognition,* University of Sydney.

Giddens, A. (1994). *The constitution of society.* Berkeley: University of California.

Grome, A., Crandall, B., Rasmussen, L., & Wolters, H. M. (2012). *Incorporating Army design methodology into army operations.* U.S. Army Research Institute for the Behavioral and Social Sciences, Research Report number 1954.

Hatch, M. J. (1997). *Organization theory: Modern, symbolic, and postmodern perspectives.* Oxford: Oxford University.

Jason, G. (2001). *Critical thinking.* San Diego State University's Wadsworth Thomson Learning.

Laszlo, E. (1996). *The systems view of the world.* New Jersey: Hampton Press.

Lewis, M. W., & Grimes, A. J. (1999). Metatriangulation. *Academy of Management Review, 24*(4).

Lichtenstein, B. (2000). Generative knowledge and self-organized learning. *Journal of Management Inquiry, 9*(1).

Morgan, G. (2006). *Images of organization* (Updated ed.). Thousand Oaks, CA: Sage.

Naveh, S. (2004). *In pursuit of military excellence.* New York, NY: Frank Cass Publishers.

Naveh, S. Schneider, J., & Challans, T. (2009). *The structure of operational revolution.* Booz, Allen, Hamilton.

Osinga, F. P. (2007). *Science, strategy and war.* Oxon: Routledge.

Paparone, C. (2013). *The sociology of military science: Prospects for postinstitutional military design.* NY: Bloomsbury.

Porter, P. (2009). *Military orientalism.* New York, NY: Columbia University Press.

PBS televised interview (2001). Interview with General Thomas Montgomery (Ret). *Ambush in Mogadishu*. Retrieved September 24, 2013, from: http://www.pbs. org/wgbh/pages/frontline/shows/ambush/interviews/montgomery.html

Ranciere, J. (1991). *The ignorant schoolmaster*. K. Ross. (Trans.). California: Stanford University Press.

Rittel, H., & Webber, M. (1973). Dilemmas in a general theory of planning. *Policy Sciences, 4*(2).

Ritzer, G. (1975). *Sociology: A multiple paradigm science*. Boston: Allyn and Bacon.

Romjue, J. (1984) *From active defense to AirLand Battle: The development of Army doctrine, 1973–1982*. Berkeley: University of California Libraries.

Romjue, J. (1997). *American army doctrine for the post-cold war*. Fort Monroe: Military History Office, United States Army Training and Doctrine Command.

Rosenau, J. N., & Czempiel, E. (Eds.). (1992). *Governance without government*. Cambridge: Cambridge University Press.

Ryan, A. (2009). The foundation for an adaptive approach. *Australian Army Journal for the Profession of Arms, 4*(3).

Sanders, T. I., & McCabe, J. A. (2003). *The use of complexity science*. Washington, DC: Washington Center for Complexity and Public Policy.

Schön, D. A. (1963). Champions for radical new inventions. *Harvard Business Review*.

Schön, D. A., & Rein, M. (1994). *Frame reflection*. New York: Basics Books.

Simons, A. (2010). *Got vision? Unity of vision in policy and strategy*. Carlisle: US Army Strategic Studies Institute.

Summers, H. G. (1982). *On strategy: A critical analysis of the Vietnam War*. New York, NY: Dell.

U.S. Army Training and Doctrine Command (2008). *Commander's Appreciation and Campaign Design*. Fort Monroe : TRADOC Pamphlet 525-5-500.

U.S. Army Training and Doctrine Command (2011). *Army Force Management Model* Fort Belvoir. Retrieved September 24, 2013, from: http://www.afms1.belvoir. army.mil/files/primers/AFMMJan2011V2.pdf

U.S. Army Training and Doctrine Command (2012). *ADRP 5-0: The Operations Process*. Washington D.C.: Headquarters, Department of the Army.

U.S. Army Training and Doctrine Command, G2 (2012), *Operational Environments to 2028*. Washington D.C.: Headquarters, Department of the Army.

Wass de Czege, H. (2010). The logic and method of collaborative design. *Small Wars Journal*. Retrieved September 7, 2013, from http://smallwarsjournal.com/ jrnl/art/the-logic-and-method-of-collaborative-design.

Weinberg, G. (1982). *Rethinking systems analysis and design*. Boston: Little, Brown and Company.

Zweibelson, B. (2013). Building another tower of babel. *Small Wars Journal*. Retrieved July 15, 2013, from: http://smallwarsjournal.com/jrnl/art/ building-another-tower-of-babel

CHAPTER 11

THANK YOU FOR YOUR SERVICE!

A Design Intervention for Transitioning Soldiers

Skip Rowland
Banner Cross Inc.

ABSTRACT

This chapter is a case study of a design intervention for soldiers transitioning out of the U.S. Army and re-entering civilian life. The purpose of the intervention is to assist soldiers with bridging the gap between military and civilian cultures and, ultimately, help them design and develop a competitive job search strategy within a comprehensive career development plan. Millennial managers and mid-career professionals are encouraged to think about the impact of returning veterans on training and hiring practices, diversity and inclusion issues, and business and societal responses to this wave of newly unemployed workers.

The intervention takes a design approach to engage transitioning soldiers, their family members, Army transition assistance services, and the community as a whole to enlist support as well as raise awareness of the multidimensional

Millennial Spring: Designing the Future of Organizations, pages 225–240
Copyright © 2014 by Information Age Publishing
All rights of reproduction in any form reserved.

transformation required of soldiers as they exit the military and re-enter civilian life. Central to the intervention is the facilitation of a shift in mindset and behavior from a mindset appropriate for military life to one that enables successful navigation of civilian culture. Design elements of the intervention are the use of a design salon environment to create a collaborative, cohort-based learning space, and the adoption of an Entrepreneurial Mindset[sm] to successfully execute the required personal and professional transformation. An informal case study structure is used to describe the main components of the intervention and share preliminary findings.

INTRODUCTION

This chapter is an informal case study of a design intervention that engaged U.S. Army soldiers transitioning into civilian life in the Pacific Northwest Region of the United States. As a millennial manager or mid-career professional, you are asked to consider the unique experiences of those making the transition from military to civilian life, but also to look for similarities and shared experiences with your own career transitions. As thousands of military service members enter the civilian job market over the coming years, having empathy for the challenges they face as they navigate this problematic transition will be important to your ability to effectively evaluate, hire, and integrate veterans into your work teams. Addressing this issue, from a management perspective, is a question of diversity and inclusion. From a design perspective, this chapter offers you a look into a program designed to change mindsets toward a more entrepreneurial approach to career development and competitive job search strategies. Ultimately, problems of planning and leading under uncertain conditions are as common to military leaders as they are to civilian, but awareness of the unique experiences and mindsets of military service personnel will be a factor in your ability to manage and lead your organization in the new millennium.

Every month, thousands of soldiers are leaving the United States military as a result of the drawdown of troops from the wars in Iraq and Afghanistan. This is putting a strain on an already tight job market and is increasing the number of applications for unemployment benefits at a time of congressional gridlock and extreme budget pressures. Many who are exiting from military service will face long-term unemployment and many have severe medical challenges. The flow of veterans into the domestic economy has been occurring for at least the last three years and the impact of this influx will continue to be a factor in the foreseeable future. On a more personal note, veterans are not only facing a very difficult job search experience, but they are making significant social, economic, psychological and emotional adjustments in the process. The multiple and interrelated problems created by this situation certainly deserve the designation of "wicked."

When you have been in a group conversation with service members and civilians, you may have heard, or perhaps even said yourself, "Thank you for your service!" It has become commonplace and is said almost without thinking about its multilayered meaning or how the service member may feel about this phrase. On the surface, it is intended to convey respect and honor for service members' contributions, but underneath there is hollowness to this cliché that may point to a growing gap between our military and civilian communities. We acknowledge, with these words, that we have benefitted from their service, but in the same breath we may be sending a subliminal message that we are not full partners in helping them survive in this competitive, global, civilian job market, which may well be as foreign to them as going into a firefight in Fallujah would be to us.

This chapter will hopefully encourage you to develop a holistic approach to competitive job enrichment and comprehensive career development planning. As a current or future manager, you will be considering veterans for positions in your organization and, if hired, you will need to help smooth their integration into your work culture. Being empathetic about their personal and professional journeys will allow you to make informed decisions. Being sensitized to the complexity of issues surrounding this scenario allows you to be proactive with strategies that are helpful to your organization and to be able to anticipate the emergent social implications that will result from the increasing patterns of interactions occurring at military/civilian interfaces. Consider your personal participation at these interfaces. How will you handle yourself? Will you educate yourself on the situation and have enough empathy for those making the journey to help them with the transition? Are you open to multiple points of view or will you just say, "Thank you for your service?"

As this chapter is structured as an informal case study, it begins with setting the context of the overall problem space and a brief description of the setting. The intervention employed to help bridge the gap between military and civilian communities is then described and key findings and suggestions for further study are offered. The voices of soldiers are used to bring reality to the discussion and although these are anonymous vignettes, they come truthfully from stories shared by soldiers within the program design salons and from personal interactions with transitioning soldiers.

THE INTERVENTION

Context

The flow of veterans out of military service is impacting all of the major civilian institutions (educational, economic, political, and social). The educational system is adjusting to enable the massive retraining required to ensure veterans have the skills required to be job ready. The economic system is absorbing thousands of newly unemployed who will need housing and all the related services. The political system is weighing the whole issue of placing our troops in foreign lands and what our future posture in this regard should be. This includes legislation, such as the Vow to Hire Heroes Act (H.R. 674) (2011) that has been passed to attempt to mitigate veteran unemployment. The social system is dealing with the impact of thousands of severely wounded veterans and the societal question that asks why we sent them in harm's way. Recent statistics tell a grim story.

There are nearly 900,000 unemployed veterans in the United States—a staggering figure. The latest Department of Labor unemployment report shows that in October 2011, the average unemployment rate among all veterans was 7.7% and 12.1% for veterans returning from Iraq and Afghanistan. Equally troubling, veterans between the ages of 35 and 64, the group with the highest financial obligations and the fewest available VA education and training options, continue to make up nearly two-thirds of all unemployed veterans. Overall, nearly 1 in 12 of our nation's heroes can't find a job to support their family, don't have an income that provides stability, and don't have work that provides them with the confidence and pride that is so critical to their transition home (House Committee on Veterans Affairs, 2012).State agencies, counties, cities, small towns, corporations, small businesses, and not-for-profit organizations are all alerted to this burgeoning problem and are contributing to the effort to assist transitioning service members leaving the military. Across our region, companies such as Alaska Airlines, Amazon, Microsoft, Comcast, IBM, Starbucks, The Boeing Company, and many others are sponsoring a wide variety of programs designed to contribute to the easing of the transition challenge to service members. The State of Washington, county governing bodies, and city councils, especially those near the several large military bases in this region, such as Joint Base Lewis-McChord, are all voicing concern, but solutions are elusive. There are many dimensions to this problem space. The military and civilian communities are both looking for ways to solve the complex problem of how best to re-integrate veterans into the civilian workforce; however, there is no consensus about how to do this and certainly no holistic design that unites the efforts of the military and civilian communities to address the entire problem space. It is a situation that is especially acute for U.S. Army Infantry veterans because they represent the largest force within the military. It is this sector of the military that is the focus of this case study.

The problem space that is addressed in this chapter is multidimensional and its complexities cannot be fully addressed here. The chapter describes one intervention designed to address one area of the problem, that is the career transition of soldiers from military to civilian work cultures and to let you hear the voices of some of the soldiers as they experience the transition. Perspectives are offered that you can weigh with your own experiences of career-related transitioning that may help you empathize with the soldiers' stories. There will be other elements of the discourse that may be foreign to you and you are asked to consider these and, perhaps, inquire about them when you next have an opportunity to interact with veterans in social or workplace contexts.

The Setting

The setting includes the unique geography of the Pacific Northwest portion of the United States and the State of Washington, the Puget Sound area, the Seattle metropolitan business area, and Joint Base Lewis-McChord, a combined Army and Air Force installation formed in February, 2010. It is a major center of U.S. military power and prestige, mobilization, and training, and is also a major center for the maintenance and promotion of the military's most cherished traditions.

The Puget Sound region is a melting pot, with one of the most ethnically diverse populations in the United States. The region is somewhat isolated, wealthy and economically competitive, highly educated, with a government focused on making the region an innovation hub. Knowledge work is the main employment engine, with computer- and bio-tech as well as alternative energy-related startup activity as growth industries. Relationships here are grown over time. Many friendships and circles of concern can be traced back to elementary, middle, and high school, and especially tight family relationships.

Design Focus of the Intervention

"How we deal with emergent problematic conditions depends on the quality of the approaches we use and try to implement. These approaches depend more on our philosophy and 'world view' than on science and technology" (Pourdehnad, Wexler, & Wilson, 2010, p. 1). The enablement of a life transition, such as that being navigated by soldiers re-entering civilian society, starts with recognition of the need for change in thinking patterns (Burke, 1994, p. 60). For veterans that transition in thinking is from a world view or mindset that they developed within the culture of the military and the mindset they will need to be successful in civilian life, and particularly the mindset they will need to be competitive in the regional knowledge economy career marketplace. This focus on thinking about thinking is rare. It is not a subject in most universities, and it is seldom found in [Western] culture [but] everything you do, want, or feel is influenced by your thinking (Paul & Elder, 2014, p. 41). This focus on how one structures thought, the "self-talk" one employs with oneself, therefore, is as critical to you as it is to the successful transition of soldiers in the program. The practice of the suspension of belief, the willingness to forego ones habits of mind from the old way of thinking and adopt a new direction of thought is hard work. "One of the hallmarks of a critical thinker is the disposition to change one's mind when given a good reason to. Skilled thinkers actually want to change their thinking when they discover even better thinking . . . yet comparatively few people are reasonable by this definition; few are willing to change their minds once they are set" (p. 44). The challenge that soldiers face is the radical transformation in their thinking required due to their deep, and frequently long-term, commitment to the military culture. One perspective you may want to think about when you are considering hiring a veteran is the strength of mind it takes to unlearn the habits of a powerful culture, like the military, and trace new habits of mind to accommodate civilian cultural norms. Fran, a Sergeant First Class, who had joined the Army at age 18, explained it this way:

> Imagine yourself as you were in high school—young and inexperienced. Now imagine you went right from high school into a job in a foreign country with a culture totally different from the one you grew up in. The job you have is dangerous. You work with hazardous materials and threats cannot be easily predicted. You are rigorously trained; you drill and practice skills constantly, but the environment you work in is unstable and complex and you are always on guard. You don't speak the language of this place and there are few cultural cues that you recognize or can rely on. It is difficult to know who is friend or foe and, in fact, the friend of today may well be the foe tomorrow. So you stay close to those who came with you to this place, because you understand them, they speak your language, they share your fears, and ultimately they become your brothers and sisters and you protect one another. Together, you form a cocoon of trust and shared experience and you live in that cocoon with them for years. It's almost like living in a parallel universe from the world you left behind you. Now imagine, suddenly, you are picked up and parachuted back into that other universe you used to live in. You are alone and the landmarks have changed. The world you have re-entered has become the foreign land. The skills you built to be successful and survive in that other universe don't seem to work here. The cocoon that had protected you and the family you were a part of is gone and you are as vulnerable as you were when you left; the strength you have relied on does not have the same value in this different reality.

This story is not every soldier's story. However, it is important for you to consider it because you, as a millennial manager or mid-career professional, will be challenged to have empathy for the "fit-gap" that veterans are facing as they re-enter civilian life. It can seem like they are parachuting into a parallel universe as they work to make sense of how things have changed. If you think about this story, you will remember that they have already navigated a much more difficult journey than you can, perhaps, imagine, and that the value proposition they can often bring to the workplace is their ability to operate with precision in highly stressful environments, where ambiguity is the order of the day, and where flexibility is required to survive, which, interestingly, describes attributes of the modern workplace.

In 2010, the U.S. Army Training and Doctrine Command published a new Army Operations Field Manual (FM 5-0) that described the environment within which the Army conducts operations as "characterized by four clear trends: growing uncertainty, rapid change, increased competitiveness, and greater decentralization" (p. v). In his Foreword, Commanding General Martin Dempsey introduced design into Army doctrine with this publication in order to better prepare field commanders "to confront a variety of complex problems, most of which will include myriad interdependent variables and all of which will include a human dimension" (p. v). He instructed commanders to "apply design to understand before entering the visualize, describe, direct, lead, and assess cycle [and explained that] combining design with the military decision making process provides Army leaders with a more comprehensive approach to problem solving under conditions of complexity and uncertainty" (p. v). The modern global business environment could easily be substituted in this description for Army Operations.

Although design as a discipline is only beginning to be discussed and exercised within the Army command structure, as well articulated in the previous chapter of this book, it is now a formal element in the Army's Training and Doctrine Command. This means it will be taught in Army classrooms and there is an expectation of its inclusion in operational plans. It is not much different within the organizational cultures of business; it is still an emergent technology, but as this book demonstrates, is becoming embedded in organizational systems across industry sectors and will be vital for you to have in your professional toolbox.

STRUCTURE OF THE DESIGN INTERVENTION

The design intervention to assist soldiers with bridging the gap between military and civilian cultures was a 12-week program of three hours each week. It was integrated into the Army's transition assistance program at Joint Base Lewis-McChord and consisted of job shadowing visits to businesses, mentoring relationships, and community engagement. The fourth major component of the intervention was the design salons where the Entrepreneurial

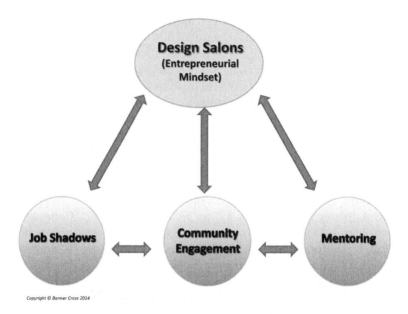

Copyright © Banner Cross 2014

Figure 11.1 Military transition learning system framework.

Mindset[sm] was taught and collaboratively discussed as an enabler to shifting soldiers' mindsets from military to civilian life. The components and their interrelationships are shown in Figure 11.1.

Job Shadowing Visits

The Job Shadow experience was designed to offer soldiers structured visits to a variety of corporate and small business work cultures. The purposes for the visits included (a) developing multiple perspectives of civilian cultures and how they operate; (b) encouraging interactive planning between transitioning service members and participating businesses; (c) learning how the different business sectors in the region interact to economically develop the region, and most importantly; and (d) understanding and discussing how the different business sectors contributed to a concept of regional economic wholeness. The job shadow visits also helped soldiers identify work cultures that appealed to them and helped them design a competitive job search process and a comprehensive career development plan to meet the requirements of those companies of interest. Soldiers were given up to five "shadow opportunities" inside participating businesses and corporations throughout the Puget Sound region.

Mentor Support

A mentoring component was designed to partner transitioning soldiers with mentors within a Puget Sound corporation or organization in order to help them with (a) the cultural challenges of transitioning to a civilian career; (b) developing a professional civilian brand; (c) positioning that brand in the Puget Sound regional workforce; and (d) helping service members develop a network of professionals in the business community. Mentor support was important to soldiers because it provided them with intimate information on workplace protocols and helped them develop professional networks.

Community Engagement

The purposes for this component of the program were to (a) broaden community engagement and awareness of the challenges facing transitioning service members; (b) encourage regional businesses and corporations to support the program with job shadow visits and mentor engagements; (c) encourage civic and community groups to promote recognition of transitioning military service members; and (d) form an ad hoc committee or "transition taskforce" to identify and obtain resources to work together on the problem and understand it as a regional issue. Specifically, the taskforce is intended to spearhead the engagement of the community. The scope of their work includes identifying public- and private-sector agencies and companies that will support job shadow visits and assisting with finding mentors. This element of the program is the least mature, but is, perhaps, the most vital piece. The extent to which the community engages will ultimately determine the fate of transitioning soldiers. But the community needs to recognize that the military must have a presence in the regional economic plan, and that military service members are, in fact, community members. That recognition, when it comes, will drive community action to accommodate the reality of today's military service member transition challenges.

Design Salons

Design salons are teaching and learning sessions that guide the composition of the principal deliverable of the program, the soldier's career development plan. The salons offer a cross-section of points of view that consider similarities and differences between military and civilian work cultures, reviews past experiences, and identifies possible points of leverage. The salons operated in this intervention as "design spaces" in which military traditions and contexts were suspended and a safe environment was created that allowed for two-way interpersonal communication. A relaxed, nonhierarchical environment was established where soldiers could openly deal with their career transition strengths and weaknesses and collaboratively identify strategies for managing opportunities and threats. The salons had a cohort-style structure (participants who entered together attended together and got to know one another). This increased their willingness to share their personal reflections. A special salon for military spouses was designed to include them and surface their unique challenges that were emerging from the transition experience. The learning content of the salons was based on the Entrepreneurial Mindset^{sm} as described in the next section and the practical application of those entrepreneurial concepts to the elements of the intervention.

Learning System

This learning system is a transformational leadership development model called The Entrepreneurial Mindset^{sm}. It is based on whole systems design principles, which consider the people, processes, structures, information, and technologies and their patterns of interaction and emergent properties, within a defined scope. The scope within which this model is applied in this context is the design intervention for transitioning soldiers at Joint Base Lewis-McChord. Dynamic systems elements involved in the design scope are political, social, educational, and economic. Through the creative elements of the design process, such as understanding mental models and a growth mindset, soldiers begin to see themselves as designers of their futures. As their design path becomes more their own, they *develop enhanced self-confidence* and assume the role of "CEO of their own careers." Within this expanded vision, soldiers develop a comprehensive career transition plan and competitive job search strategy.

The Entrepreneurial Mindset^{sm} is a holistic change agenda that consists of seven elements that operate synergistically: Preparation of the imagination provides the foundation of understanding of the power of the human mind and its visualization engine and how that enables the creation of motivating visions, which require the skill of gap-closing communications to share motivating visions with stakeholders. Game-changing attitudes fuel the formation of audacious goals, which set the change agenda that is implemented through maximization of effort. Collaborative evaluation enables the assessment of the value of actions and the iterative process of continuous improvement based on feedback. Each element is described below:

Preparation of the Imagination establishes the human brain as the foundation for learning and mindset change, reviews how the mind operates as the natural center of composition and creativity and engages participant imaginations with compelling growth-oriented thinking. It describes the imagination as a mental process that connects many levels of human consciousness. It encourages singleness of purpose, the organization and direction of the mind to a "definite chief aim" (Hill, 1928 p. 61), and the use of mind/brain psychology to overcome fear, build positive relationships, and design dreams. It helps participants understand the big picture and "little snapshots" of their professional and private lives with a balanced focus intended to help them fully consider regional employment patterns and not make critical judgments too quickly.

Motivating Visions challenge participants to build mental models to vividly illustrate their chosen career path. The idea of lifestyle design is discussed and participants experiment with their own models based on the assumptions and constraints that their career direction allows. Participants develop three mental models in their imagination and write them down on paper. The first mental model is the development of a clear picture of the participant's current lifestyle. The second mental model is a vision of the participant's desired or dream future lifestyle and the third mental model is a moving picture of the experiences, skills and abilities they would need to develop in order to live their dream lifestyle. The building of relevant mental models is a skill anyone can learn to give themselves a competitive edge for business and career transition and advancement.

Gap-Closing Communications focuses participants on ways to "close interpersonal gaps and reduce interpersonal conflict that may occur due to individual differences and limited resources" (Myers & Myers, 1982, p. 242). In communications between transitioning soldiers and business professionals, a cross-cultural communications gap can be created by the simple use of the phrase "Thank you for your service!" The gap occurs due to differing perspectives. The transitioning soldier may perceive a business person as a potential employer and is looking for something concrete in the way of an employment lead. The business person may just be having a conversation and the tendency is to use the phrase as a positive end to the conversation. The business person thinks they are ending the conversation on a positive note. Not so for the soldier who is in a different culture from the business professional, feels a desperate need to secure employment, and is not sophisticated in competitive business protocols. The competitive job search requires job seekers to build multiple networks and deal with multiple perspectives. Multiple perspectives can lead to the mis-interpretation of behavior and interpersonal communications problems. This section of the learning system helps participants prepare for such challenges by practicing courageous conversations, which are difficult discussions about where job opportunities exist, how to gather proprietary information about the opportunities, and how to position themselves for those opportunities.

Game-Changing Attitudes help soldiers understand why beliefs and attitudes influence perceptions and behavior. This area of learning shows why entrepreneurial habits and attitudes are important to a career transition strategy. Developing an entrepreneurial mindset focuses participants on how their personal beliefs and assumptions impact their ability to perceive employment from an employer's perspective. The methods employed disrupt soldiers' comfort zones and allows for the development of new habits and attitudes more relevant to the civilian workplace. The guiding belief is that if people understand how their attitudes are shaped, they can effectively design an attitude-shaping process that truly reflects their career aspirations. (Design attitudes are discussed in Chapter 13).

Audacious Goals with Measurable Objectives teaches the value of setting "stretch goals" that represent the achievement of success in career development, while at the same time recognizes current realities. The goal-setting process begins with a clear mental model of the desired lifestyle result, moves through an iterative process of career development research and self-evaluation, and ultimately leads to an integration of lifestyle choices and supportive career possibilities. This learning area establishes goals as targets for the mind that must include a budget and a deadline.

Maximization of Effort demonstrates how to synthesize the learning elements previously described and organizes their collective energy through a comprehensive career development plan. "When you organize your faculties...and direct them toward the attainment of a definite purpose in life, you then take advantage of the cooperative or accumulative principle out of which power is developed, which is called organized effort" (Hill, 1928, p. 81). Organized effort can then be focused on specific goals, which in this case were the development of a professional civilian brand and competitively positioning that brand in the marketplace.

Collaborative Evaluation encourages the consistent and persistent collection of data and measurement of productivity to evaluate and reframe the plan. This puts the focus on the ongoing process of observation and judgment that make iterative reframing and intentional change possible. Soldiers were taught to seek data inputs from multiple perspectives; perform regular qualitative and quantitative evaluation of job search strategies; and monitoring of the regional economy, the job market, and their position in that job market to look for patterns and themes. The most important element of this section of the intervention was the experience of collaborative dialogue within the learning cohort, and between the soldiers and their mentors and with other members of their business development network.

FINDINGS AND SUGGESTIONS

- These findings are my personal reflections from the two years experience I have had working with soldiers at one Army base in one region of the country. The findings are not generalizable but hopefully point to areas that are worthy of further study and, perhaps, to formal research that would yield more generalizable results. Managers and professionals of the future will face new workplace diversity and inclusion issues not yet on their radar. Perhaps we should acknowledge that the military is a culture distinct from civilian culture and consider the military as a group that should be a focus of diversity programs. It would follow that managers should educate themselves on the culture of the military and study the transition military service members make in their re-entry to civilian life in order to effectively integrate veterans into their teams. Managers and professionals should also be prepared for miscommunications with veterans due to cultural programming. Enhanced business communications between the two cultures begin when managers seek out soldiers to gain a more diverse perspective.
- There is an apparent cultural disconnection between civilian communities and the military. The reason this is important is that it is this gap that military service members must bridge as they re-enter civilian life. And, in a very real sense, they are crossing that bridge alone, one-by-one, in the thousands, and although many programs are springing up monthly to "help," there is little military/civilian engagement in dialog about holistic solutions to close that gap. In a very real sense, we as a society are focusing on the symptom (the need for soldiers to transition out of the military and find new careers) and few are discussing the reason why that transition is so hard. Many are saying "Thank you for your service!" but few are having the courageous conversations required to close that gap.
- A design approach and methods appear appropriate for dealing with complex social problems such as the business and social relationship between local civilian communities and transitioning military service members, but these relationships will take time to be fully recognized and appreciated. The design salons that introduced the discussion of having an entrepreneurial mindset required open discussion of strengths and weaknesses and collaboration across rank structure. This was uncomfortable for many participants and the viability of holding design salons on a military base may need to be rethought. The following quotes from soldiers tell the story. One soldier said, "It is hard for me to express my true feelings with senior officers in the room. It is much easier for me to let the officers

238 ■ S. ROWLAND

lead the discussion and have the final word." Another soldier said, "I am afraid that if I speak up in here, there will be penalties later on!" And an officer said, "It may be better if you separate senior officers and NCOs from the lower ranking troops. That separation may improve communications for all!" Even if it is required to separate the ranks during the initial transition learning program, I suggest that before the end of the program transitioning service members of all ranks come together for structured dialog.

- The context of a program like this shapes the outcome. In this case, the competitive regional job market and the sheer number of transitioning soldiers suggest that a well-resourced and integrated skills development and job acquisition program is needed.

- This design intervention appears to have been a "disruptive technology" (Christensen 1997, p. xviii). The disruption was an unintended consequence of the intervention. "We cannot know for certain that what we design is what ought to be designed. We cannot know what the unintended consequences of a design will be and we cannot know ahead of time the full systemic effects of a design implementation" (Nelson & Stolterman, 2003, p. 10). The military community appears to have some reservations around a focus on the development of an entrepreneurial mindset. One of the lower ranking soldiers in my class said, "They are not going to let you continue to do this. You are stepping on too many toes!" Even though innovation is becoming a part of many corporate visions, the concepts that come with an entrepreneurial mindset are still considered new technology by many. The inclusion of design in the military operations manual mentioned in this chapter indicates that design is entering the lexicon of the military and that the direction is shifting toward more self-directed and flexible decision making paradigms, at least at the command level of the military. With the lower ranks of the military, it may take longer for these ideas to be accepted. The soldiers ultimately appreciated the disruptive nature of the intervention and worked to move from a more fixed-mindset to a growth-mindset. Once the minds of the soldiers began to shift; however, I noticed that it was difficult for them to go back to the boundaries of their military perspective. They were caught between mindsets and it is a very uncomfortable place to live.

CONCLUSION

Hopefully, this chapter has motivated you to familiarize yourself with the challenges facing those exiting from military service and you see opportunities for your own professional growth from the material shared. As the military reduction-in-force process moves forward, more and more servicemen and women will find their way into the civilian work force. Millennial managers and midcareer professionals will have to assess, hire, and integrate veterans into their existing work teams. Empathy for the service member will be critically important, and balancing that empathy with the needs of civilian organization will be challenging.

Business managers and leaders of the future may also find themselves interfacing with the community on behalf of transitioning service members, perhaps as a part of future diversity and inclusion initiatives. Since employment problems impact the community as a whole, both the public and private sectors will call on its leaders and managers to provide guidance and engage with the military. Those managers and leaders who plan in advance for this challenge will have a competitive advantage over those who do not.

As a millennial manager or mid-career professional, you will have the fortunate or unfortunate challenge of making business decisions at a time when the world is learning to see itself as an integrated whole. Learning to look at problems from a whole systems design perspective is a learning agenda that can help you integrate diverse cultures and, in effect, manage, in positive ways, human perceptions of difference. Civilian institutions, given the proper leadership and appropriate orientation, will take on the challenge of publicly recognizing and discussing the role our military plays in the design of our economy. A design approach has proven to accommodate complexity and seeks to understand the systemic interconnections that are holding problems in place. With this approach, we can all do more for our servicemen and women than say "Thank you for your service!"

REFERENCES

112th Congress of the U.S. (2011). Vow to Hire Heroes Act (H.R. 674). Downloaded from http://www.gpo.gov/fdsys/pkg/BILLS-112hr674enr/pdf/BILLS-112hr674enr.pdf on 2/5/14.

Burke, W. (1994). *Organizational development: A process of learning and changing*. Reading: MA, Addison-Wesley.

Christensen, C. (1997). *The innovator's dilemma*. New York, Harpers Business Essentials.

Department of the Army, Headquarters (2010). *Army Field Manual (FM 5-0)*. U.S. Army Training and Doctrine Command, Washington, D.C.

Hill, N. (1928). *The law of success*. New York, Penguin Group.

House Committee on Veterans Affairs (2012). Vow to Hire Heroes Act. Statistics downloaded from http://veterans.house.gov/vow on 2/5/14.

Myers, M., & Myers, G. (1982). *Managing by communication: An organizational approach*. New York, McGraw Hill.

Nelson, H. G., & Stolterman, E. (2003). *The design way: Intentional change in an un-predictable world*. Englewood Cliffs, NJ: Educational Technology Publishers.

Paul, R., & Elder, L. (2014). Learning the art of critical thinking. *Rotman Magazine,* Winter, pp. 40–45, Rotman School of Management, University of Toronto Press.

Pourdehnad, J., Wexler, E. R., & Wilson, D. V. (2010). *Systems and design thinking: A conceptual framework for their integration*. 2010 International Society for the Systems Sciences 55th Annual Conference, University of Hull, Hull: UK.

PART V

A DECADE OF PROGRESS

CHAPTER 12

A CALL FOR STRONG DESIGN LANGUAGE[1]

Fred Collopy
Richard Boland
Case Western Reserve University

ABSTRACT

Editorial: We are delighted that the leaders of the 2004 workshop that produced *Managing as Designing* share their assessment of the progress in creating the designs for the future.

INTRODUCTION

Everything about an organization is designed. Most often the design approach taken by managers is to borrow a design already in use by another business, function or location—one that addresses most of the issues and seems to work well enough for them. When the search for this design is conducted systematically it is often elevated to the status of a "best practice". That approach has some obvious problems. For one thing, "best practice" connotes that it will be very difficult to find a better solution, since the best is already being used, which inhibits further work on creating a better design. In addition, because each organization is a unique assemblage of

Millennial Spring: Designing the Future of Organizations, pages 243–250
Copyright © 2014 by Information Age Publishing
All rights of reproduction in any form reserved.

assets, capabilities and environmental dependencies, any "best practice" consensus is unlikely to align well with those of your company or situation, or that of any particular organization.

This is a small but telling example of how the language of management can aspire to be strong and capable, yet unwittingly end up being weak and limiting. In this short chapter, we urge you to bring the strong skepticism of youth to the weak language of your elders by keeping a strong design attitude alive in your managing practice. Your design attitude is a way of perceiving the world around you and a readiness to act that you can hold close and use every day. A design attitude is simply the expectation that "Nothing is as good as it can be, and everything could be other than it is". It is a simple but bold way of being oriented toward the world around you that is hyper sensitive to the poor designs that we encounter in the world at large and in our own organizations. We first observed the design attitude in our study of the architectural practice of Frank O. Gehry. He is a remarkable artist, struggling every day to make the built environment more functional, but in the strongest, broadest sense—meaning aesthetically, emotionally, spiritually, culturally, morally, inspirationally, and humanly functional—not just economically functional. Economic functionality is a part of any design effort, but alone it is a weak part. Alone, it is a minimal goal and will result in designs that are default designs, unsatisfying and not worthy of us as full human beings.

In a presentation at the Weatherhead School for the opening of his Peter B. Lewis building in 2002, Gehry expressed his exasperation at the absence of a design attitude evidenced in our built environment. We all know, he said, that economic goals are only a part of what we hope to create as leaders, whether architects, managers, politicians or craftsmen. We all know that.

Why then is there so much mediocrity in our landscape? Why then doesn't the world at large realize it? I'd say 98.5% of buildings are mediocre—I call them *buildings* because I wouldn't even list most of them as architecture.

We don't have a simple formula for reversing the quickness with which our leaders in business, finance, healthcare or government move to the selection of default, minimally functional designs, or continue to accept ones that are in place. Instead, we wish to propose that it takes a design attitude, applied relentlessly to an evaluative perception of your surroundings along with the strength of conviction to challenge the existing ways in which things are designed, to work towards a more desirable organizational world.

The Design Attitude and the Manager

The design attitude that we saw in the day to day work of Frank Gehry can be incorporated into the work of all designers, including the work practices of all managers. In describing managers, Peter Drucker opined that they are the people who create the conditions in which other people work. This characterization of managers suggests that they should posses a design attitude and engage in design practices. And fortunately, design attitudes and practices are not the province of a rarified talent. Rather, as the great artist and designer Karl Gerstner asserted, design "must not be misunderstood as an activity reserved to artists. It is the privilege of all people everywhere." We would add that for managers it is both a privilege and a responsibility. For the millennial manager it is also a special opportunity to bring fresh eyes, fresh sensibilities and a fresh belief that "We can do better than that!"

This volume and others that have been written in the past ten years testify to the fact that managers are turning to design in their struggle with the problems and the challenges they face with the complex systems at the core of their businesses. Our interest over several decades has been to better understand what lessons-learned from professional designers transfer into the realms inhabited by managers, so as to aid them in these important responsibilities.

The emerging generation of managers are the right people to engage in the designing of new business forms, products and processes because they bring relevant analytic skills that are a necessary complement to designing complex business systems, along with the deep sense that everywhere they look, things are not as desirable as they could be and should be. And managers have shown growing interest in this combination of skills. For example, a 2009–2010 study in which IBM interviewed 1541 CEOs in 60 countries and 33 industries included this among its conclusions: "The degree of difficulty CEOs anticipate, based on the swirl of complexity has brought them to an inflection point". Asked to prioritize the three most important leadership qualities in the new economic environment, *creativity* was the one they selected more than any other choice. Fully 60% of the interviewees included it among their three qualities, more than integrity, global thinking, influence or openness.

But creativity in what? In problem solving, certainly, but also in problem finding—meaning the ability to make an abductive leap of conceiving ways that things could be other than they are, and could be more functional in a strong, broad sense. The millennial generation of managers are positioned at that inflection point, with a strong skeptic's belief that things are not as good as they could be, coupled with the commitment to design all aspects of today's organizations to be more broadly functional. This includes the design of more fully functional products, services, processes, policies, structures, cultures, climates and strategies. Nothing is exempt from an assessment and intervention with a design attitude.

Strong Design

Design is an incredibly strong, powerful word. As a verb, it denotes *the giving of form to an idea.* What could be more closely tied to the unique generative capability of human beings? It is the capability to take our thoughts, desires or concepts for improving our conditions of life and giving them real form in the world—as physical artifacts, practices or methods. But design is also one of those words that slips easily from a highly meaningful engagement to a weak, everyday, inconsequential label. From the giving of form to an idea, which places design in the highest levels of human aspirations and accomplishments, involved in the shaping and reshaping of our world, it can easily slip to the mundane and weak meaning of being "stylish", mere "fashion", or even a plan to deceive.

Worse, design can, and often does, shift from an active striving of our cognitive and emotional intelligence engaged in the risky creation of something new, to a passive, merely contemplative activity located in our own heads. When the meaning of 'design' shifts from describing a peak of human achievement to something closer to the lowest level of sentient being, energy spent on adorning oneself, we have no one to blame but ourselves. The meanings of words are the raw materials of our cognition and our situation awareness. The way we use them is a choice we make. The question is how do we make those choices and are we even aware that we do?

Design Thinking versus Design Making

Design is the making of things. It is an active engagement in creating the new, projecting our thoughts and ideas into the world around us, and making them visible for all. *Design thinking,* a phrase that is becoming a sort of shorthand for creative approaches to problem solving, is quite the opposite. Its focus on thinking points us to its role as a hidden, inner process that is apart from action. The very phrase lends added support to an already widely held notion that mind and body, thinking and action, are separate entities, which is an idea that supports dualistic thinking in general. In turn, the world and its problems come to be seen as causal chains between separate entities instead of as mutually constitutive aspects of one and the same being.

Consider the following: if, instead of the common practice of using *decision maker* as a description of a manager's work, we used the phrase decision thinking? As you consider that phrase, we will share our sense of how things would play out. First, who comes to mind when we say decision thinker? To us, it sounds like a staff person, relegated to the back office, a mere advisor to the real manager. In contrast, a decision maker takes action based on a higher order analysis of data, some of which may be provided by a decision thinker. The decision thinker does not make strong moves or make real changes, he or she just thinks about it. Ironically, this image of the manager/leader as a thinker instead of a doer is congenial to us because it comes down to us from the Greek philosophers and their distinction between thinking and acting. For them, thinking was a higher order activity than doing. The leader who thinks is the master planner who has minions carry out the required actions. Action is below the leader and unworthy of their contemplative role. It is too close to labor and too far from the ideal realm of thought, especially the most prized form of abstract thought.

That voice of dualism is behind many of our present ills. The voice of privilege is the voice of abstraction and analysis, distant from the dirty work of making, doing and acting productively. We know, however, how destructive and wrong the voice of dualism generally is; how it can lead us to emphasize mind over matter, spirit and contemplation over physical toil. We know that our mind and our body are one, inseparable and densely interactive. Maya Lin, the designer of the Vietnam Veterans Memorial puts this unity nicely when she asserts: "I think with my hands." It is in the manipulation of objects, especially models of design alternatives, that we do our best thinking. It is the moment when our abstract contemplation and ideas for improvement become a real and important part of our work and our world.

Psychiatrist Iain McGilchrist sees the roots of this dualism as intrinsic to the very structure of the human brain. "If one had to encapsulate the principal differences in the experience mediated by the two hemispheres, their two modes of being, one could put it like this: the world of the left hemisphere, dependent on denotative language and abstraction, yields clarity and power to manipulate things that are known, fixed, static, isolated, decontextualized, explicit, disembodied, general in nature, but ultimately lifeless. The right hemisphere, by contrast, yields a world of individual, changing, evolving, interconnected, implicit, incarnate, living beings within the context of the lived world, but in the nature of things never fully graspable, always imperfectly known—and to this world it exists in a relationship of care. The knowledge that is mediated by the left hemisphere is knowledge within a closed system . . . It can never really 'break out' to know anything new." In his book, *The Master and His Emissary,* he chronicles how these two ways of being have been engaged in something like a battle, noting that the left hemisphere has played a dominant role in our recent history, a period during which we have also seen the dominance of a purely analytic mindset in much of management practice.

In what follows, we discuss some of our observations of design work and of designing at work, hoping that you will develop an appreciation for design as action. We will draw upon both our own experience as designers and empirical research in which we and our colleagues have gotten to know the work of designers. We use *design* and *designing* as verbs instead of noun and adjective. We will focus on the term "design thinking" which slowly seems to be displacing the use of design as a making. We will recount how the word system, which had such promise 60 years ago, slowly degenerated to "systems thinking", and we will explore the consequence for losing the power that the word system once promised. When Churchman used the term "system approach" it was a strong and exciting way of engaging real problems in the world head on. It meant a remaking of the familiar into a new reality through a dialectic process of acting, building, testing, reflecting, and . . . acting.

How Design Matters

Among the most direct ways of investigating design is to engage in it. In this section we explore what our own experiences have taught us about how the problems of systems are shaping design practices. In doing so we are engaged in the search for Kenneth Boulding's second order laws of design, who said when speaking of the generalist: "If laws in his eyes are good, laws about laws are delicious and are most praiseworthy objects of search."

One of us has spent his adult life writing computer programs. In doing so, he has engaged continuously as a designer. Indeed, we would argue that virtually all computer programmers are designers (in practice if not also in theory). This because only rarely do actual programmers execute on a specification that has been designed by others. Rather, they use whatever time is available to them to iterate in the interest of such unspecified values as elegance, maintainability, extensibility, and if they have any ego at all, originality. This is the best explanation for the asymmetry of prediction errors for the completion time of programming projects. Any available time (and then some) can be used to iterate upon the design.

From years designing commercial consumer software (including *The Desk Organizer*, published by Warner Communications), we have learned that design solutions must align with the ecosystems for which a product is designed and that each ecosystem has its own feel, style and rules. We learned that craft matters greatly and that craft is a matter of localized knowledge. And we learned that products and their businesses must be co-designed and allowed to co-evolve.

Together we designed another large software system, *Business Animator*. *Business Animator* uses a graphic input-process-output model, the cycle model, to represent key ratios of a business. The dynamics of supply chains, income statements and balance sheets were depicted with movement and color and without any numerals. In experimental settings, subjects, even experienced financial analysts, who used these entirely graphical representations were able to make better decisions about the health of businesses than when they used conventional representations, including spreadsheets and traditional time series charts.

Our work with *Business Animator* taught us once again how much craft counts and that some of the features that were most critical to success were not among those in our early specification. Instead they emerged in the act of designing, as students experimented with building and validating models.

In September, 2013, the Weatherhead School launched the first Department of Design & Innovation in any business school in the world. This was the result of a decade long initiative aimed at embedding design practices into the education of management undergraduates, MBAs and executives. The initiative has been grounded in ongoing conversation with designers, research on specific design practices, and experimentation with a variety of pedagogical approaches to embedding design in management education.

While working with others to build an awareness of design as a foundational skills for managing large systems we learned additional lessons. The act of bootstrapping an initiative is in itself confidence building. Frank Gehry refers to the importance of providing handrails to help people understand a new take on or application of an idea. The nature of an ecosystem both constrains and enables possibilities. Because CWRU has neither a design nor an architecture school, for example, we were able to achieve things that Universities with design or architecture schools could not. At the same time the fact that our new building placed us on the same piece of land as the Cleveland Institute of Art shifted the kinds of collaboration that took place.

Given the questions that most interest businesses today, we are usually concerned not with the necessary but with the contingent, as Herbert Simon puts it, not with how things are but with how they might be. The matters we face are so complex that we need strategies for controlling process complexity. Design is a useful approach because it is, as Gregory Gargarian observed, provisional, reflexive, evaluative and confidence building.

What You Can Do About It

What can you, as the future of managing, do given the important role that design plays in acting? We suggest that there are at least three things. The first is to understand and develop a personal design attitude. The next is to learn a few design practices that work for you; tricks that will enable you to step back from the problem at hand, to reframe issues, to see the world's problems in the context of the systems that generate them, to generate your own alternatives, to rapidly prototype and test solutions, and to select among then with the aid of your deepest aesthetic and ethical intuitions. Finally, given that you create the conditions in which others work (and do business), you should consider making workplaces arenas in which engaging in design is first safe, and then respected and facilitated.

The essence of these three is what we call strong design. The manager who practices strong design has a strong design attitude, with strong design skills and a strong sense of design space. With those three qualities, the new generation of managers can indeed do better than we have done. You can look upon the organizational world you are inheriting and say: "we can do better than that!"

PART VI

DEBRIEFING

CHAPTER 13

THE ART OF CONDUCTING DYNAMIC EMERGENCE

Michael N. Erickson
Boeing Company

ABSTRACT

And just when we thought buzzword bingo was finally starting to fade away . . .

I tend to dislike buzzwords and fads, but I work in an industry filled with them.

Among the buzzwords and seemingly faddish ideas that cross our paths is that set of concepts around "self organization" (beginning with Lorenz's "Strange Attractor," and then the "Dynamic Emergence Principle") that comes out of chaos theory.

These words are all attempts to describe the same recurring notion that some view as a little too crazy to think about, and certainly not a thing tied to anything useful in temporal reality.

But increasingly there are people (such as myself), who take these ideas seriously and have come to depend on them. The key questions are "How is that possible?" and "Whatever were you thinking when you entered THAT space?"

The following is an attempt to untangle the thinking around how we might properly apply emergence in what we laughingly call "the real world."

Millennial Spring: Designing the Future of Organizations, pages 253–278
Copyright © 2014 by Information Age Publishing
All rights of reproduction in any form reserved.
253

INTRODUCTION

This recurring notion of self organization, as described by quantum physicists as a force for good, seems simply too good to be true. Aside from the built-in and very threatening notion that chaos is inherently and totally uncontrollable, thus more likely destructive than anything else, it's often hard to even consider thinking seriously about how self organization would be in any way practical in regard to the subject of design (engineering, organizational, or even product design).

It would be easy to assign it to that stack of so-called "magic bullet" business solutions that we've endured wave after wave of over the last decades. The ideas around standardized design techniques and processes, which come either out of business management schools or from the minds of well known business gurus, and so often designated as "flavor of the month," have had their time in the spotlight over the years, each claiming to be "just the thing" needed to make everything straightforward, innovative, and wonderful.

Sadly, it rarely works out that way, and for most of us dealing with them just makes us tired.

Rather than becoming liberated, highly productive, and innovative, we find ourselves caught yet again in another collection of the same complex, bureaucratic, and cumbersome tasks that we complained so bitterly about and that cost us so dearly in our previous experiences.

And while the accusation can often be made that we "didn't do it right" (and "we" probably didn't), we do eventually come around to the notion that one man's solution isn't necessarily "every person's solution," so the search goes on.

Thus it tends to be that we look on terms like "dynamic emergence" with considerable cynicism, and why not?

Before we can talk about it, though, we need to take a close look at how design is typically handled. These approaches are quite reflective of "the magic bullet" approach to design. You'll quickly see that there are some very good reasons to reconsider our use of them.

WHAT THE WORLD WANTS

Creativity is that thing that if properly evoked leads to innovation, and EVERYONE wants innovation. The western world has bet its entire economy on the requirement for unending innovation. . . .

Yet while they say they want innovation, and they say they want breakthrough thinking, they remain entrenched in polarized debates; locked forever, it would seem, in a black-and-white view of reality. One could say that we are very often driven to mediocrity, and it's killing us.

EVERYONE knows that creative people are at the very least unpredictable. Maybe even a little bit crazy. They are NOT safe people, because you can't predict how they will respond, much less when they will deliver the results and especially the innovations we assign them to find.

This represents chaos and unearths a basic worry that chaos will lead to destruction, NOT the innovative solution.

SO WHY, THEN, IS CREATIVITY AND INNOVATIVE DESIGN SO HARD?

The problem with creativity is that it doesn't come when we call. It doesn't show up when the project plan requires it, and it doesn't follow any structure or order, much less "the rules." Creative people are generally very random access and unstructured in their thinking, and almost always out of step with whatever staged activities are going on at the time. They are radical, ask scary questions, make messes, disrupt meetings, work at odd hours, and scare the hell out of the managers.

Because to create, they don't follow process; they simply create. Rather than control, they "cope," or "ride the wave," or "follow the energy". . . .

And in doing that, they reset all future possibilities for everyone around them.

Creativity (and creative people) tend to be chaotic, and there are few things in this world scarier than chaos.

IT'S ABOUT THINKING!

Did you notice how many times I mentioned in the previous paragraphs the word "thinking"? I contend that it's how we think about things that controls our design effectiveness.

To talk about thinking, I often use the analogy of the iceberg. While what appears on the surface can seem daunting, it's what is below the surface and out of sight that contains the greater mass and most often is of the greatest concern.

We all remember the Titanic don't we? It was what wasn't seen that destroyed that era's most striking example of engineering design genius. The example of the Titanic stands to this day as evidence that if we don't pay attention to all the available information, we run dire risks. There is nothing that we create that is unsinkable.

LOOKING AT THE SURFACE—WHAT WE DO IN PURSUIT OF "SAFE" DESIGN PRACTICE

Over the centuries there has evolved a more or less standard set of ways charted to get design to happen. These "ways" of operating are not necessarily creative or innovative. They have never been offered as anything ideal, in fact it would seem that these are the same ways of operating that the "magic bullets" offered by our management consultants and gurus are designed to solve.

The only thing they seem to really offer is their familiarity, and some sense of being in some way "safe." Safe from what, no one is sure, but to depart too far from them is a sure way to elicit panic, and if *Dune* author Frank Herbert was correct, fear truly is "the mind killer" (Herbert, 1990.

So here's an incomplete list:

Design by Recipe

There is a common idea that "if you've done it once, you should be able to do it again...just follow the process you did last time." I say this works only if you're trying to do exactly the same thing you did last time. If you want something new, you can't follow a canned process; you have to look beyond it.

Design by Accountants

Working in aerospace I often found myself assigned by my management to invest some time in a project or idea, only to have someone in an accounting or finance capacity stop me in my tracks and overrule my management.

I also watched in horror as a set of design concepts that were really quite elegant were gutted, or the design process was radically redirected because of demands by the accounting groups to conform to a set of cost requirements that really made no sense from an engineering perspective. I began to wonder if there was a special type of "magical thinking" inherent to the accounting world that was not tolerant of the explorations required to develop a quality design, and instead expected a certain amount of walking on water, or possibly mind reading to be standard practice.

What typically came out of these kinds of design efforts was never terribly creative or revolutionary, since emphasis on safety and cost minimization drove the design teams backward to mediocrity, or to a tendency to "go off half-cocked" or to make commitments before anyone is truly ready.

I find design by accountants to be just sad.

Design by Requirements

So in talking to any engineer or software developer, you will sooner or later encounter the notion of "requirements" and the absolute belief that good "requirements gathering" is of critical importance in good design.

This might be true if you're gathering the right requirements, but what if that's not all that's going on?

While getting accurate system and process descriptions is of critical importance, making them "requirements" for success, thinking only in terms of requirements is often far too austere a mind-set to deliver the depth and versatility required by high quality design.

I consider design by requirements to be "design to minimums"; while it produces workable designs, it so often lacks that intangible something.

Design by Combat

In my early days as a beginning systems analyst, I drew a cartoon reminiscent of a fight scene in the old Andy Capp newspaper comic strip (something like this one here).

I titled it *Rough Modeling*, because I found myself in a room with 40 people, all shouting at each other in acronyms, and this cartoon seemed to perfectly capture the combative effort it seemed to take to capture and define

the various system, data, and process elements we needed to put into the models.

This "cross dysfunctional" design team had been brought together to define the business process and the top-level system requirements for a major component of the manufacturing production system of a new airplane program, and it was very rough going.

It became evident that the viewpoints held by the various technical experts in the room were both profoundly colored and individual. Our subject matter experts (SME's) would come into the room with most often a very defensive stance, prepared to debate and to defend their view.

Each participant would come in with his or her 10 top things and descend in assault on the other people's 10 things, so that when all the dust cleared there would be only a few major things left standing on each side . . .

Resulting in a VERY anemic design.

But our cultural mental model insists on this combative approach (we call it "logical debate") under the belief that this is the safest approach to sorting out truth, failing of course to notice that this approach is "exclusive" (polarized—"I'm right while YOU are wrong") and that we exclude anything that we can possibly defeat through debate, rather than an inclusive design paradigm.

The outcome of this kind of design, and the profound costs of this exclusive approach, are finally showing that it NO LONGER serves us.

GETTING BELOW THE SURFACE—LOOKING AT THE PREVAILING MENTAL MODELS

While the surface "ways" of working are driven by our thinking, we have behind the scenes (and below the obvious surface of things) a shared set of fundamental ideas about how reality works that we hope work for us rather than against us. The term "mental model" is sometimes used to describe these accepted ways of thinking.

The concept of the mental model was presented to the world by MIT professor Peter Senge in his book *The Fifth Discipline* (1990) as a concept that could be described in a nutshell as "things we believe that pass for 'normal'" . . .

The difference between a mental model and a belief isn't much. What is important about them is that by seeing them clearly we can recognize the difference between what we believe and what happens in actual practice.

Among the driving concepts/beliefs/mental models of our civilization is a deeply held belief pertaining to perceived safety or "structure" . . . which ostensibly leads to "order" that can be managed (or better said, "controlled").

The biggest driver in a corporation, after the bottom line, is the need to control.

Part of this has to do with efforts to take appropriate responsibility (to, for example, customers, stock holders, and reg- u l a -tory agencies), driving companies to operate at a "process" level. The rest, we must admit, results in the addiction to having power over people, driven by a belief (or mental model) that equates strong leadership with structure, order, and safety.

Control is considered to be essential to an ordered civilization.

This point of view results in a perception that "civilized" design is something done under the constraint of structure, with a strong emphasis on the use of repeatable processes, again noting the proliferation in recent years of a broad assortment of formulae and "best practice" aimed at the delivery of quality, good design, successful implementation, and whatnot.

Yet, we face increasing disconnects in our control-centric design efforts that continue to result i n frightening economic effects that reflect poor design decisions, the missing of design milestones, the seemingly erratic progression to the design effort, or the higher risk this adds to design programs, not to mention the ongoing popularity of Dilbert cartoons mocking the entire engineering design mind-set. This suggests that the current mental model driving our design methodologies has a fundamental gap (or two).

Among those gaps is the tendency toward polarized thinking, the notion that there is one right way (right or wrong, black or white, correct or inferior) and that the debate or "control over" mind-set will help you find it.

Due to this, it would seem that in western corporate culture the "control over" model is among the strongest of the mental models in operation, resulting in most of the action (including design action) occurring for the most part at the surface of things.

That surface could be defined as "immediate stuff," like budget, schedule, agenda. It's the day-to-day "stuff" needed to run a business.

Hostage to our own thoughts..

But I believe this surface is an illusion that keeps us both in a reactive state and managing in reverse, basing our plans on past successes and previous ideas. Not much creativity, much less "envisioning," happens, because you can look at only the surface of things.

Thus the "solution concepts" (what I've been calling magic bullets) generally offered by business schools and business researchers only provide an appearance of progress, while under the surface, where the real issues are, little attention is paid and little improvement felt.

The search for the Holy Grail of a good repeatable process that can be depended upon to produce consistent results drives all manner of business acrobatics, where the symptoms of a problem are attacked rather forcefully while deeper system or process drivers are ignored (or flat rejected). Very often it's the very old ideas—things about invoking curiosity and taking the time to really ask questions and dig out the answers—that, if examined and seen for what they are, could be employed to provide design solutions that are far more complete and satisfying and that actually work over the long term.

AND THEN THERE'S THE MYTH OF "NORMAL"

If mental models are more deeply examined, then the idea of "normal" (which is the basis for most mental models) must be reexamined.

One could say that the notion of "normal" is a key belief driving standardization. It includes the mythology that there exists a certain "core" standard-issue human (or way to be human). This "being normal" is interesting when we see that what has been regarded as "normal" depends upon the historical era and what part of the world one lives in, and that it often changes, sometimes minute to minute.

While being countered to a certain degree by ideas arising from the diversity movement, where it's the differences in people that is seen as the fountain for creative insight, the myth of normal, where we are regarded as being "all the same" to a certain statistical degree, still prevails and attempts to establish a standardized way of working, and worse, a standardized way of thinking. While the myth supports a certain limited utility in the support of "factory flow," as found in the Lean manufacturing concept, the end result is a not very creative place, and also certainly not a very adaptable place to be, because there is always someone standing over you telling you that you need to change your thinking to be with the program.

It makes me wonder where the line is that divides our general similarity (that thing we call being normal) from mediocrity?

THE FEAR AND CONTROL MIND-SET

We live under the aforementioned mental model, which leads us to believe that we need to control the design process so that we are locked down in patterns of the repeatable process that will supposedly save us, and that standardization is the best and "safest" bet.

We pathologically fear failure, not realizing that experiencing failure is a key element to achieving success.

Well, at least we talk a good fight...if one prefers not to notice the record of irrational and incredibly blood soaked world wars, economic panics, disease epidemics, and natural disasters that have given clear evidence that humanity is, at best, only marginally managing to stay ahead of the game.

"OUT OF THE BOX" AND OTHER TRITE TERMINOLOGY

It would really help us if creative and innovative design emerged on its own, spontaneously, and came and rescued us. What seems to be really needed is a third option, another choice, a deeper solution, one that allows us to truly "think outside the box"...

So there's a conflict here. Our business leadership wants the innovation that leads to good design and good outcomes, but the conditions that generate innovative thinking are seen as too "dangerous" to be permitted, since they can't be controlled, monitored, or planned.

In the context of design, the old ideas around what constitutes innovation and creativity run counter to what business claims to want, because it demands CONTROL over creativity.

Creativity on demand doesn't work. Instead, what happens is this...

DIVING EVEN DEEPER—EXAMINING THE TROUBLESOME
ELEMENT OF "PERSONAL EXPERIENCE"

So once we admit that canned design practices have limitations, leave things out, or don't go far enough, and that maybe we should examine the models that frame our thinking, we begin to be able to rely more and more on our experience and on our "gut." We start to trust ourselves just a little bit. We risk a little and begin to step outside our known boxes, and TRUST in ourselves (which is very important) becomes possible—when it comes to properly wielding dynamic emergence.

We live in a culture that tends to be suspicious of personal experience. Personal experience is often viewed as highly individual and verges on being "antiscience."

We've built an entire culture around how we utilize peer reviews and agreements to establish the parameters of our perceptions of reality. The entire point of an academic degree is to establish the affirmation (from outside of ourselves) that we can do a thing we claim to be educated to do.

Having proven that we are capable of earning a degree by following an established curriculum that always reflects the standard practice and is supported by statistical evidence (this being called science), we tend to ignore other kinds of learning, especially those anomalous personal notions and those frightening divergent viewpoints considered to be random flashes of temporary insanity, or at best, hallucinations.

Yet without mentioning in any detail the time (sometimes literally years of time) spent cleaning up after and dealing with the residual effects of the "great and wonderful" buzz infested solutions unleashed by my executive leadership, I've quite naturally developed some clear ideas, backed up by experiences of how to actually get good design to happen, and I don't think they are hallucinations. I also don't believe they are "anti" science.

I would submit that there is more going on in the design arena than meets the eye, that it is a "multivariate" problem that must be addressed using a broader set of tools.

On the one hand we have our latest and greatest "buzzed" idea that promises great things, usually derived by means of a "best practice" someone has discovered, or based on a business school theory where statistical data has shown certain measurable outcomes in the test state, and on the other we end up with a history of struggle and frustration.

And all the while there is the ongoing state of corporate thinking, that regardless of what supposed visionary motivation is presented, when push comes to shove (and there's always something economic going on that creates both push and shove) the demand for bottom line profits and adherence to predetermined schedule trump all other concerns, and as a result whatever gains are made in an organization fall by the wayside. What could have been "good design," and possibly even "great design," ends up becoming "just barely enough design to get by...yet again.

This situation raises a question. If these magic bullet ideas are so great (and especially this notion of self organization), then why don't they overcome and supersede this tendency to regress?

To understand that, we must examine the environment that we have, and ask ourselves what it is that gets us into this situation in the first place.

SO WHY DO I BELIEVE THIS STUFF?—PERSONAL STORIES THAT EVOLVED MY OWN DESIGN CONSCIOUSNESS

So my own awareness of creativity in action started with personal experiences. I learned by watching my father, and later my grandfather, build things....

They were both carpenters, from different generations, but who shared a certain suspicion of the academics, as well as a thoughtfulness and practicality in their work, that would be considered in the present day a kind of hacker mind-set, where they would try things, explore possibilities, and make startling discoveries.

Though very functional in their approach, even their most "whacked together" creations had a certain rough-hewn elegance and even beauty to them. While not terribly prolific, both made wonderful things.

My father was also a cranky kind of old guy who did not work very well with others (no team player this one), so he most often was assigned to go off by himself, to do those detailed and finicky tasks most didn't want to do, like hanging doors.

He was also the go-to guy when there were difficult problems to solve, and after he got past his obligatory laughter and covert mockery of the engineers and their "elaborate contraptions" he would then proceed to lay out the perfectly square building foundations with a tape measure, four stakes, and some string. He could cut long compound curves in long planks with only a hatchet and could handcraft wood, leather, and bone into a wide array of very practical, though maybe also quirky and counterintuitive, artifacts, tools, toys and more.

Sadly, although he spent the last 12 years of his carpentry career caring for Tom Sawyer Island in the Anaheim Disneyland park, where his skills really fit, he totally failed to recognize the level of artistry he applied to his create and build process. He did not consider himself a "designer," because for him it was not a remarkable thing that the doors he installed were perfectly balanced, and the windows he repaired never stuck. He was, after all, "just" an old-time carpenter.

To his credit, I see few carpenters (or any other artisan) with such offhanded and practical "design" artistry these days.

DESIGN THEORY

The first thing I was ever exposed to that could be considered design theory came to me by way of a garage shop inventor/imagineer called Ed Knight, who taught me what he called the "total concept" design approach.

I used to have long conversations with him over about a seven-year period where, using hot rods from the 1950s as a good working example, he would describe how modifications made to one part of the machine (such as the engine or transmission) made in isolation would utterly fail to improve the vehicle performance, because the entire system was then put out of balance. . . .

I later encountered this concept, with other descriptive language attached, in *The Fifth Discipline* (Senge, 1990), where the very Zen notion that to touch one element was to touch them all, requiring a lot of careful thought to be invested in what Ed Knight would have called a total concept design.

The core lesson for me was that good design was first and foremost about thinking.

But as a rule thinking, at least the way we do it, isn't just a free form exploration that yields unfettered creativity. Thinking is often colored or constrained by a great many things that Peter Senge described as mental models. These models put parameters around what we believe is possible, or practical, and quite often our mental models become traps all their own.

One of the most common mental models, the "control over" model, demands that there be structure to thinking, which generally results in a recommended set of commonly prescribed steps. While this is generally viewed as the safest approach, it also is a deeply constrained approach that automatically precludes many possibilities.

MY FIRST ENCOUNTER WITH DYNAMIC EMERGENCE

I was present in the room when a certain technical expert "saw" for the first time something he'd been talking about for a very long time, and what he saw required him to make a fundamental shift in his thinking.

It began in a corporate graphic support office. This expert had come in with a request for a graphic designer to construct a conceptual model of the expert's understanding of a certain technology, and the model was to be included in a technical document with 50 copies made by morning for an important design project.

The graphic designer took the most amazing set of accurate "design" requirements I've ever seen, and produced the model faithfully in every detail, to the specifications given. A team of designers, writers, editors, and reprographics personnel worked all night to have the document ready by morning.

And I was there in the morning when this technical expert came to pick up his documents and saw his model for the first time.

Upon seeing it he expressed shock at what he saw.

"This is wrong!" he told the designer. "But this is what you said, and here is the list of requirements to prove it!" she retorted. "But this leaves out this whole other thing," he stated flatly. "Well you did not speak of that other thing, so how could I put it in?" she demanded.

"Well this is wrong and must be done over."

I must at this point make it clear that this graphic designer did an exemplary job of turning the words of this technical expert into the picture he said he wanted.

I must also point out that what really happened was truly amazing, and is of critical importance to the subject WE are exploring in this chapter.

When the technical expert "saw" what he had "said," for the first time a different part of his brain was activated (that part that sees, relates, and conceptualizes), and he discovered holes in his thinking.

THE CONCEPT THAT CAME OUT OF THESE EXPERIENCES...

He could NOT have previously understood what he now discovered, because most of the knowledge he was attempting to express was collected and ordered by the predominate thinking process that our culture supports (that side of the brain that processes information in the linear and logical fashion).

The moment he saw what he had said for the first time was the moment he understood what he had previously "known" from a broader perspective than he ever had before.

Sadly, I think he may have missed the significance of this event, because he left the office grumbling that his schedule was now at risk. The graphic designer was also upset, because she now had to do the job over again at considerable cost. . . .

But the significance of what happened was NOT lost on me. I saw for the first time HOW it is that visualization is critical to the thinking that goes into design (be it engineering design, artistic design, or any other kind of design).

UTILIZING THE DIFFERENT "THINKING" MODES...

As you may have noticed, I've been gradually working my way down the iceberg—deeper and deeper into the underpinnings of our way of doing design.

1. The surface could be those things we do as common practice.
2. The concept of "what we believe" (our mental models) would be that murky space just below the surface that motivates our actions.
3. Then there's that deeper space—the one we are all suspicious of— where our personal experience resides.

Now it's time to think about how we literally process our thoughts, and how this relates to the notion of "chaos" and dynamic emergence.

SEQUENTIAL DESIGN STEPS VERSUS INTEGRATIVE DESIGNER INTERPLAY...

Most of our academics emphasize "reductionist thinking." In this approach, taking things apart to understand them is considered to be the logical way to understand something. It's worked fairly well in terms of scientific examination of most of the natural world. We observe the human body for example, and notice the separate systems that operate within it, or the cardiovascular system as a separate thing from the musculoskeletal system.

These separations allow us to learn a lot of details about components and subcomponents of natural systems, resulting in an explosion of knowledge on all fronts.

The challenge with reductionist thinking, however, is that it makes it difficult to see things as a whole.

Also, since the period in history known as the Enlightenment, there has been a deep commitment to the concept of "logical thinking." Logical thinking arose out of Greek philosophy and was an object of considerable focus during the Renaissance, when great progress was made in defining its parameters, evolving in due time to Boolean logic and then into the digital machine thinking that drives our computer-centric society.

One problem with linear logic is that it is possible for logical constructs to become recursive or to "convolute." Convoluted logic in a computer is called a "software loop" and preventing them is among the lessons all software developers learn in their early training.

Logic looping in humans is more typically described as "mental illness" because, while in a loop, you cover the same ground over and over again, and as a popular saying goes, "Insanity is doing the same things over and over again, expecting different results."

The conquest of nature and the creation of great works of engineering became powerful expressions of the power and control humanity felt it had to have over its environment—if it were to truly be the masters of its (or our) own fate (remember the Titanic).

Chaos manifests itself in environmental breakdown and market upheavals that boggle the imagination.

The key point here is that regardless of how humans view their reality, they (we) are never quite in control. More often than not, great emphasis is placed on maintaining the illusion of control, the façade of structure and order, while all around us we "bail furiously" against the rising water of our own convoluted logic and short-term planning.

Copyright © 2013
Michael Erickson
All Rights Reserved

The Concept of Integrative Design

There are ways of thinking that appear chaotic, but aren't. They just don't travel in a straight line.

One emerging form of Integrative Design is called "world building" and comes out of the USC School of Cinematic Arts.

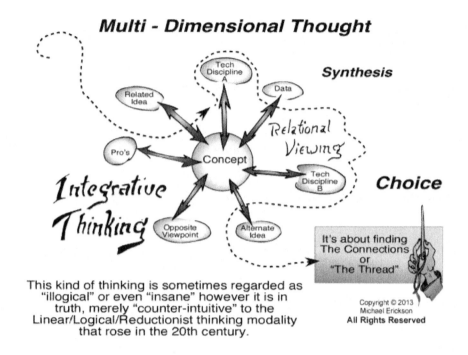

Copyright © 2013
Michael Erickson
All Rights Reserved

The Concept of Multidimensional Complexity

The essential concept is this:

We live in a "multidimensional" reality (and we always have).

We occupy an (at least) 12 dimensional "reality" that requires a multivariate and cross disciplinary collaborative approach to design solutions, where we note that:

1. The emphasis on diversity of participants (diversity in both ways of being and in skills and views into the problem) is essential to creative and innovative problem solving.
2. This approach enables discovering what advantages accessing all aspects of the "brain architecture" bring to problem solving.
3. This approach takes full advantage of the concept of "dynamic emergence," thus becomes a form of "following the energy" of more than simply emotional or enthusiastic effort.

My belief is that reality has NEVER been simple. Humanity has merely had a simplistic understanding of it. What we are beginning to wrap our minds around is that only recently have we begun to start "seeing" it...

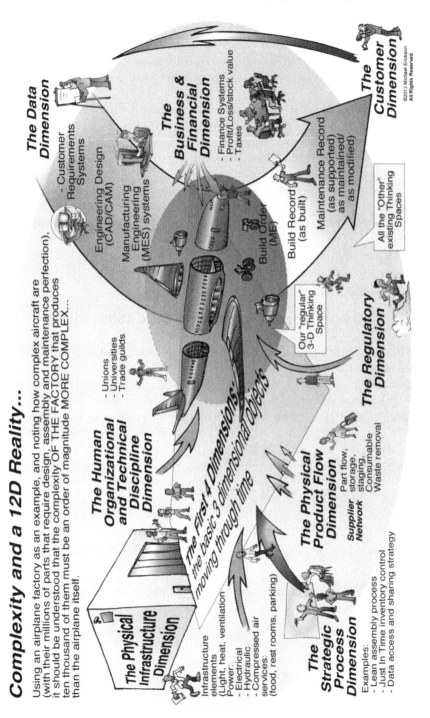

Complexity and a 12D Reality...

Using an airplane factory as an example, and noting how complex aircraft are (with their millions of parts that require design, assembly and maintenance perfection), it should be understood that the complexity OF THE FACTORY that produces ten thousand of them must be an order of magnitude MORE COMPLEX... than the airplane itself.

The Data Dimension
- Customer Requirements Systems

Engineering Design (CAD/CAM)

Manufacturing Engineering (MES) systems

The Business & Financial Dimension
- Finance Systems
- Profit/Loss/stock value
- Taxes

The Customer Dimension

Maintenance Record (as supported) as maintained/ as modified)

Build Record (as built)

Build Order (ME)

All the "Other" existing Thinking Spaces

Our "regular" 3-D Thinking Space

The Human Organizational and Technical Discipline Dimension
- Unions
- Universities
- Trade guilds

The First 4 Dimensions: the basic 3 dimensional objects moving through time

The Regulatory Dimension

The Physical Product Flow Dimension
Supplier Network
Part flow, storage, staging, Consumable Waste removal

The Physical Infrastructure Dimension
Infrastructure elements (Light, heat, ventilation Power:
- Electrical
- Hydraulic
- Compressed air services: (food, rest rooms, parking)

The Strategic Process Dimension
Examples:
- Lean assembly process
- Just In Time inventory control
- Data access and sharing strategy

©2013 Michael Erickson All Rights Reserved

So everyone agrees the world has become incredibly complex. As the horse-and-buggy era of the early twentieth century gave way to "modern" times, with its industrialization and proliferation of transportation and communication technologies, complexity seemed to be one of the persistent outcomes.

I know that because their bodies occupy only the three standard dimensions (height, width and depth), even when they account for their movement through time (the fourth dimension), people tend to think of themselves as living in only a 3D or 4D world.

So how does design play in this space?

Because most of our technical disciplines grew out of the craft and trade guilds, where the "making" tasks became fairly linear processes and then repeatable processes, we have tended to see design through a fairly narrow keyhole.

This would be an example of "the expert," who gives you the answer, in contrast to the "methodologist," who helps the answer emerge. . . .

SO NOW AT LONG LAST, LET'S TALK ABOUT CHAOS

When we hear the word "chaos" we don't typically take it as anything good. For most of us, chaos is an extremely bad thing, highly dangerous, even destructive.

The ancient Greeks may have best expressed the prevailing cultural attitude toward chaos with their legend of Zeus's defeat of the Titans. His conquest of chaos is said to have resulted in the flowering of civilization and the age of great philosophers.

Among the reasons we now call the medieval period in Europe "the dark ages" is the level of political and social chaos that reigned during that time.

The rise of the age of reason is reported by historians in tones reflecting great relief that sanity had once again returned to humanity. The general enthusiasm initiated by the likes of Leonardo da Vinci and Galileo that set the stage for the age of exploration, spurred on by the writings of Voltaire, Jefferson, and Sir Isaac Newton, led to the science and technology of the nineteenth century and the industrial revolution and culminated in the current information age.

We can hardly view the current state of civilization with too much euphoria however, considering how deeply it's embedded in the notion of nature as "the great machine," and considering the long history of determined effort to overcome the effects of chaos. . . .

While the rise of the scientific method established the role of logic and reason as among the most dependable methods for keeping the chaos at bay, it's also important to note the role of the fear lurking in the hindbrain of most "civilized" people and keeping our efforts toward control in the forefront.

An interesting side note rests in perception. The American Indians, who long existed in the western North American continent in relative harmony with the land and in what most of them have reported as a fairly orderly state of being, found the European invaders to be the source of total chaos. Not only was their life and culture uprooted, but it was replaced by concepts and ideas (about money, power, and the angry god) that they truly felt represented insanity.

So while the Europeans, who saw these "wild" people as a source of chaos, tried to force them into the structures of money and labor that represented order to them, the American Indians watched in horror as their ordered life and the resources of the land began to be consumed by the engine of enterprise. . . .

SO WHY WOULD CHAOS BE CONSIDERED A "GOOD" THING?

So if there is ONE fundamental lesson it is the increasingly obvious idea that our long history of expedient ("quick and dirty") solutions only delays the payback or backlash of poor design.

We have to do something different because, if you haven't noticed, our world is currently full of "payback."

I'm here to suggest that chaos is not, as might appear on the surface, merely a destructive force. Rather, it is a creative force. It's the essence of creative possibility, and when faced and even embraced directly it can be trusted to evoke that mysterious construct that's been called "dynamic emergence" or "self organization" that the buzz word proponents tell us is the real solution to our problems.

I have to say that I agree with them.

Wikipedia defines dynamic emergence as follows. "In philosophy, systems theory, science, and art, emergence is the way complex systems and patterns arise out of a multiplicity of relatively simple interactions. Emergence is central to the theories of integrative levels and of complex systems" (Emergence, n.d.)

My definition: Dynamic emergence is a natural process that will happen if you support it and don't fight it, in spite of the standing model that insists on a more combative approach (logical debate) that no longer serves us.

EMERGENT DESIGN CONCEPTS

When I first read Margaret Wheatley's descriptions of "the emergent dynamic" in *Leadership and the New Science* (2006), it sounded both so familiar. I found that it was about so much more than simple collaboration (cooperative effort).

Unleashing the Emergent requires a collaborative interplay, it requires us to hold the design space long enough to exhaust the known in order to open the possibility of consideration of the unknown (Wheatley, 2006).

When we hold the design space, unburden ourselves of the known, include alternative thinking skills and techniques, bring EVERYONE into equal engagement, and hold out for that final complete insight... only then do we step into the new frame of reference, that insightful spot where all the puzzle pieces fit and where the deeper understanding becomes so clear.

Debate is NOT the safest approach to the discovery of truth, because it's an "exclusive" (weeding elements out), rather than an inclusive, design paradigm.

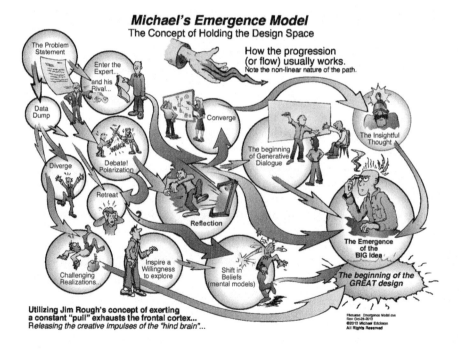

Design must go deep (beyond the obvious) and cannot be superficial or on the surface, lest it be merely an expression of yet another quick and dirty solution that won't stand up under pressure, will need to be either endlessly patched or "fixed" in order to exist at all (thus exponentially expensive). Design must be people-centered.

Jim Rough's concept is found in his book, *Society's Breakthrough!* (Rough, 2002).

THE EXPERIENCE OF LEARNING TO SEE

The Power of "The Image"

The power of the image has been known for millennia, since the very wise know that if we can picture something, we can bring it into existence.

Leonardo de Vinci, whom we all know as that famous renaissance man, "the maestro," the consummate artist, was in fact a defense contractor working for the Duke of Milan, who used art to discover "science," and who spent most of his days working up maps and siege models, not unlike many of my own aerospace engineering compadres in the company I work for.

Leonardo walked a fine line, at a time when being too much a deviant from the "standard issue" way of life could get you burned at the stake for heresy.

Thus, tight control over the image has long been held, especially by religious authorities, resulting in the many strange and challenging ideas about "the image" that confront us.

There's a standing mental model in major corporations that says something like "If it's a picture . . . it's communication" (and the graphic designers then believe they "own" that space).

They don't typically use imagery to think with.

The Concept of "Holding the Design Space" . . .

1. I feel that my key role is in "holding the design space."
2. I hold it open as long as possible.
3. I hold it from a neutral stance as long as I can.

4. I continue to pull the ambiguous into solidity as fast as the awareness of it emerges in the group, and
5. I don't relent until the discovery is made.

What I've been doing all these years is focusing on "visual thinking." There's a fundamentally different thinking process (as well as mechanisms and even the side of the brain that's engaged) in this visual thought thing that I'm working on, and while admittedly it does become "communication" eventually, where the heart of MY work is centered is in doing the Leonardo de Vinci "thing" of using art to discover science.

Holding the Design Space is...
Sweeping all those piles
of data into some
coherent order...

And doing
it again
as many times
as it takes,
for the design to emerge!

Copyright © 2013
Michael Erickson
All Rights Reserved

The greatest part of my work is in simply "catching" and visually recording thoughts in their various half baked and barely coherent initial states, when they come out of someone's mouth in their typically oblique fashion, usually as an aside from the main thing being discussed. The thing about this visual thought herding is that for the longest time what one is really dealing with is a mass of unformed and rather cluttery "stuff" that one pushes around—as if with a broom, into one pile and then into another, until coherence begins to emerge.

The key ideas I try to help people get in these conversations are about how to get the half formed pictures out of their head and onto a surface where they can look at them together, and how to help this visual "stuff" jell into something that can—in due time—become "communicate-able."

So most of what I'm intending to convey are:

1. Techniques to help people get into "draw" mode without being encumbered by the lie that it requires the "magical" thing called talent...
2. Strategies to catch rapid-fire and "buckshot" delivery of big piles of random access ideation,
3. Something about how to sweep these piles together into a coherent "picture" that WILL (eventually) become worthy of communication, but for the present time serves as a placeholder that lets the concept grow and develop in a natural (almost biological) way.

If we jump directly into "visual communication," we end up doing damage to our own engineering processes where evolution of the idea goes from very rough to rather coherent in some sort of spiral or evolutionary fashion.

One of the great risks the corporate communications standards have, regarding PowerPoint, is that an idea requires a certain level of polish to be in place in a general PowerPoint pitch, which becomes a hazard when a half-thought-through idea gets delivered in a "too pretty" condition, and is judged as "done" and "ready for prime time," which in MY business, can result in something being built badly and killing someone.

I might also point out here that when we think about who (in our Aerospace business) ultimately "owns" the pictures I must make a case for it being Engineering (not graphic designers), simply because we define our products (airplanes/military and space stuff) by means of a "picture." (An engineering drawing, whether it be a blueprint on paper or a 3D CAD/CAM representation, is STILL a picture!) So to keep our feet grounded in engineering, I submit that to hold the engineering foundation for our visualizations, be they on the back of a napkin or in a CAD simulation, they still must be owned by Engineering.

The formal communications stuff that the creative services and other corporate creative functions deal with is a subset (and a special use case) for the picture, and yes, they own that use case. But the two subject areas are rather different and need to be approached as different.

I submit that the clear distinction between visual "thought" and the finished visual explanation that is ready to communicate is a critical distinction to make.

So if we can talk about this as "visual thinking" rather than "visual communication," then I won't continue to encourage a thing that I think is highly dangerous (pictures of half-thought-through stuff that are too pretty too soon, thus too believable and therefore deadly).

So with all this, why do we need to even worry about "dynamic emergence"?

Because it's the final ingredient that brings us out of the repeating pattern, and delivers the true breakthrough design that we want so much.

I'm not so much a translator as a raw thought processor. I use art to discover science (this is where the visual element plays), and in the process a lot gets thrown away. The visuals are the transitory elements that assist in the distillation of coherence from otherwise random-access and easily forgotten bursts of insight (usually group- or team-based burstings, rummagings, and general thrashings through difficult—and usually engineering-oriented—material).

We get to the communication part eventually (sometimes waaaay eventually), so the bulk of the work is in holding the design space open for as long as possible until something recognizable pops out.

So the word "modeling" resonates. I also kind of like "visualizer," because I bring into view what people try to describe.

I think the difference between what most of you are trying to describe and what I'm talking about is in the level of clarity we work with. Whereas most of you translate somewhat digested ideas into some form where people can see it, the engineering design, factory, and business process infrastructure in my world has a lot of raw energy and dynamic interplay that takes a while to get one's mind around so we (the groups I work with) spend a lot more time in the depths before we get anything worth communicating. So I tend to think of visual modeling more in terms of shamanism (identifying and bringing into focus the energy that is currently active) than any sort of structured technical discipline.

I suspect that this view is a little too scary (or at least ambiguous) for some, but it's where I find myself and where I'm becoming increasingly comfortable and even a little bit powerful (such an unexpected aspect).

CONCLUSION

Although "design thinking" is currently in vogue (and it has a LOT to offer as reflected by some of the other write-ups in this book), I'm beginning to think the following:

It's NEVER been about finding that "one right way" or the "magic" process, to do effective and innovative design. Design thinking could fall into a definition of the best of the known processes. Maybe it is, but even when it is absent good design still happens.

When it comes to the truth about how to get good design, I think it's more about an attitude and a sense of trust. I believe that by maintaining a certain stubbornness and holding a belief that you are still on the right track, especially when things get crazy, you are getting closer to the solution.

Exhaust the Frontal Cortex...

...to release the creative impulse of the "hind brain"...

© Michael Erickson

Chaos (whether it's the perceived chaos or the real thing) is a symptom that we are near some edge that must be explored. When standing on the edge of the known world, one must decide consciously whether or not to stay and make the new discovery or run for the safety of the known.

Craziness might actually be a dependable signal (or evidence) that you are on the brink of a breakthrough, since it's generally pretty obvious at that point that NOTHING else you might have known or tried has worked, and something NEW has to be explored.

Breakthrough design requires us to take risks, trust ourselves, and hold the design space as long as it takes for the various elements to take form.

In chaos theory, the term "strange attractor," "emergence," or "self-organization" might be an unfocussed attempt to give a name to something we can't really get our minds around (the energetic dimension possibly), but the bottom line KEY technique (if you can really call it that) is to simply and stubbornly hold on to the design space until the reality you've been searching for emerges.

This might be the core "thing" that in the end, holds the greatest importance.

ACKNOWLEDGEMENT

Thanks has to be given to my teachers…both the conscious ones, who mentored me carefully, and the unconscious ones, whose rude disregard for current reality left an indelible imprint and who are the source of my caution, rebellion, and overt desire to find a better way.

REFERENCES

Attractor. (n.d.). In Wikipedia. Retrieved December 26, 2013, from http://en.wikipedia.org/wiki/Attractor

Edward Lorenz. (n.d.). In Wikipedia. Retrieved December 26, 2013, from http://en.wikipedia.org/wiki/Edward_Lorenz

Emergence. (n.d.). In Wikipedia. Retrieved December 26, 2013, from http://en.wikipedia.org/wiki/Emergence

Herbert, F. (1990). *Dune.* New York: Penguin.

Kleiner, A. (2008). *The age of heretics* (2nd ed.). San Francisco, CA: Jossey-Bass.

Rough, J. (2002). *Society's breakthrough! Releasing essential wisdom and virtue in all the people.* Bloomington, IN: AuthorHouse.

Senge, P. M. (1990). *The fifth discipline.* New York: Doubleday.

Zubizarreta, R., & Rough, J. (2002). *A manual and reader for dynamic facilitation and the choice-creating process: Evoking practical group creativity and transformation through generative dialogue.* Port Townsend, WA: Jim Rough and Associates.

ABOUT THE CONTRIBUTORS

Tomas Backström, EdD

Tomas Backström is a Swedish researcher, and a tenured full professor in innovation science and leader of the innovation management research at Mälardalen University in Eskilstuna in Sweden. He has been a researcher in the work life area since his PhD in 1996. Starting with an undergraduate degree in theoretical Physics his work has become more and more multi disciplinary and he has published articles and chapters in a wide spectrum of venues such as: *The Learning Organization, International Journal of Computer Integrated Manufacturing, Economics and Business Letters,* and *Nonlinear Dynamics, Psychology, and Life Sciences.* Two of his publications have received awards from the HR Association of Sweden.

Min Basadur, PhD

Min Basadur began developing his insights about creative thinking and problem solving at Procter & Gamble. He received three U.S. patents and created a corporate-wide innovation consulting practice. Following his award-winning doctoral research at the University of Cincinnati, he became a professor of Organizational Behavior in the Michael G. DeGroote School of Business at McMaster University. His most recent book, *The Power of Innovation,* became an instant CEO need-to-read. His research is published in top scientific journals and encyclopedias worldwide. Min founded Basadur Applied Creativity in 1981 and began developing a worldwide network of consulting and research associates applying his trademarked

Millennial Spring: Designing the Future of Organizations, pages 279–286
Copyright © 2014 by Information Age Publishing
All rights of reproduction in any form reserved.

Simplexity Thinking ™ system internationally with major organizations including: eBay, Microsoft, John Deere, Goodrich ,LG Electronics, Fox Studios, Dallas Children's Hospital, Cancer Care Ontario, Magna, Pfizer, Kimball International, PepsiCo, Frito-Lay, Toyota Canada, BASF, Procter & Gamble, Mead Johnson, Atomic Energy of Canada, USAF Aeronautical Labs, NASA, and many others.

Richard Boland, PhD

Prior to joining the Weatherhead School in 1989, Richard Boland had been Professor of Accounting at the University of Illinois at Urbana-Champaign since 1976. He has served as a visiting Professor at the UCLA Anderson Graduate School of Management, and has held the Malmsten Chair for Visiting Professors at the Gothenburg School of Economics, University of Gothenburg, Sweden. Currently, he also holds an appointment as a Fellow at the Judge Business School, University of Cambridge in the United Kingdom.

Some representative publications include, "The Process and Product of System Design", *Management Science* (1978), "Sense Making of Accounting Data", *Management Science* (1986), "Accounting and the Interpretive Act," *Accounting, Organizations and Society* (1993), "Designing Information Technology to Support Distributed Cognition", *Organization Science* (1994), "Perspective Making and Perspective Taking in Communities of Knowing", *Organization Science* (1995), "Why Shared Meanings Have No Place in Structuration Theory", *Accounting, Organizations and Society* (1996), "Knowledge Representation and Knowledge Transfer", Academy of Management Journal, (2001), *Managing as Designing*, (Stanford University Press, 2004), "Wakes of Innovation in Project Networks" *Organization Science* (2007) (which won an Academy of Management 2008 award for best published paper), Hermeneutical exegesis in information systems design and use, *Information and Organization* (2011), Organizing for Innovation in the Digitized World, *Organization Science* (2012).

Fred Collopy

Fred Collopy is the Vice Dean and a Professor of Design & Innovation at Case Western Reserve University's Weatherhead School of Management. He received his PhD from the Wharton School of the University of Pennsylvania. He has published over fifty articles, reviews and notes, on forecasting, technology, strategy and the application of design ideas to management. He co-edited the book *Managing as Designing*, which was published by Stanford University Press in 2004. Fred has designed several large systems including *The Desk Organizer* (the first product in the personal assistant category when it was published in 1982 by Warner Software), *Rule-Based Forecasting* (an expert system for selecting among alternative business forecasting models),

Imager (an instrument for playing abstract visual images like musicians play abstract sounds), and *Business Animator* (an interactive representation of accounting and financial information).

Peter Coughlan, PhD

Peter is a core faculty member with the Organization Systems Renewal program at Pinchot University in Seattle, WA. He is also an affiliate with the Institute for the Future in Palo Alto, CA. Peter has spent the last 20 years in the field of design, working at IDEO in product and service development, as well as systems and organizational design. Over the years his work has shifted toward application of design methodologies (such as observation and prototyping) to complex organizational challenges. He has been working with Colleen Ponto over the last three years to create a process that integrates systems thinking and design thinking in service of solving complex organizational challenges.

Andrea Cifor

Andrea Cifor is a 6 Sigma renegade—abandoning projects toward developing lighter, faster methodologies that lead organizations to successful change. Throw Information Architecture into that mix and she defines strategies where data is the asset not just a byproduct of computing. With a mantra of "good BI should make decisions not drive them" her secret obsession lies in the optimization and automation of complex logic in response to business intelligence and data. Andrea currently lives in Atlanta, GA working in the data quality capacity at a digital agency, she is an IASA CITA-P Information Architect as well as Six Sigma and LEAN certified.

Michael N. Erickson

Michael N. Erickson is a Visual Practitioner, of the BCA Processes and Tools organization, Requirements & Architecture Integration Team (RAIT), based at the Boeing Company factory in Everett Washington. Born in Alaska, and having lived the majority of his life in Washington State, his life experience ranges from the very rudimentary (rough carpentry, felling trees, hand forged tools, ice climbing) to extreme technology (system, concept and process) design. Essentially a "systems analyst who draws" he is primarily known for large system visualizations or models, (could be described as "Journey maps", "Vision, Concept of Operation, System, Process pictures") and contextual storyboards. A heavy user of cartoon art, which allows him to safely depict touchy worrisome issues pertaining to the human condition in the context of organizational stress and the complexity of large scale system integration.

Miriam Grace, PhD

Miriam Grace is a practicing whole systems designer and thought leader in innovation at The Boeing Company. Miriam is an Enterprise Architect in support of Sales, Business Development, and Marketing Systems. She is responsible for all technical solutions and the business technology strategy for Boeing's customer-facing functions. Miriam is a member of the Boeing Technical Fellowship that represents the 1% of Boeing's technical workforce that has been honored for their significant contributions and track record of breakthrough achievements. Miriam is a leader, teacher, and mentor in the use of Design Thinking as a mindset and method for the transformation of organizational culture to accelerate innovation and competitive differentiation. She is a Fellow of IASA Global, an international professional organization of practicing architects with established chapters in 35 countries, where she chairs the Curriculum Committee of the Board of Education and reviews and approves all architecture courseware that is delivered on-line and in person to the over 3000 members worldwide. Miriam holds a master's degree in whole systems design and a Ph.D. in Leadership and Design from Antioch University.

George (Bear) Graen, PhD

G Bear is a well-known authority on research and practice in the people operation functions of business organizations globally. His professional training includes the Industrial and Organizational Psychology, Doctoral program at the University of Minnesota, and faculty appointments at the University of Illinois, Keio University in Tokyo, the University of Cincinnati, Nagoya Imperial University in Nagoya, the Hong Kong University of Science and Technology, the University of Louisiana (Cajun) and Donghua University (Textile) in Shanghai. His Butterfly theory (*aka* LMX Theory) of people operations is internationally recognized as the best data-based explanation of human collaboration in organizational design, production and service teams (*Research Review on Executive Development of Leadership*, 2013, Oxford University Press).

Deborah E. Gibbons, PhD

Deborah holds a bachelor's degree in psychology from the University of Washington, and masters and doctoral degrees in organizational behavior and theory, with a statistics minor, from Carnegie Mellon University. As an associate professor at the Naval Postgraduate School, she teaches system dynamics, team-building, leadership, intercultural collaboration, and a variety of managerial topics. She specializes in integration of social psychological principles with data analysis and system modeling. Her research addresses networks that support collaboration, knowledge-sharing, and community-building; humanitarian aid and disaster response; concurrent effects of personality and social context on cognition and behavior; and diffusion of information, attitudes, and behaviors in multi-cultural environments.

James K. (Jim) Hazy, MBA, EdD

Jim is a tenured Full Professor at Adelphi University in New York State in the United States with nine years of academic experience in research and teaching. Prior to joining the university, he worked in business for 25 years including serving as an executive at AT&T and Ernst & Young, LLP. Having published over 40 articles and chapters in venues such as *The Leadership Quarterly*, his work has won numerous awards including the Academy of Management Best Paper Award and the prestigious Bender Body-of-Work Research Award for the period 2009–2011. In 2010, he co-wrote the book, *Complexity and the Nexus of Leadership: Leveraging Nonlinear Science to Create Ecologies of Innovation*, which achieved Amazon Top-100 status in Management Science in both hardback and paperback editions. He has also co-edited the books, *Complex Systems Leadership Theory* and *Complexity Science and Social Entrepreneurship: Adding Social Value through Systems Thinking*. His teaching experience includes: entrepreneurship, strategy, operations and leadership delivered to undergraduate and graduate students as well as executives.

Marcus Jahnke, PhD

Marcus is a Swedish researcher who holds an MFA in design (2005) from HDK, the School of Design and Crafts at the University of Gothenburg, and a BSc in Innovation Engineering (1994) from The University of Halmstad. He is currently a senior lecturer and researcher in design at HDK where he conducts research and lectures, for example at the master program in Business & Design. His current research interest concerns the intersection between design practice and innovation, which was also the topic of his recently finished PhD study. Marcus has also studied gender and design issues through experimental research approaches and in collaboration with several companies and authorities. Other issues of interest concerns sustainability and social innovation. Marcus has a practice background from the automotive and building sectors where he worked with environmental management systems in the 90s and early 00s in the Volvo Group as project coordinator and at NCC Construction, a division within the Nordic building company NCC, as environmental manager.

Grant Martin

Grant Martin is a Lieutenant-Colonel in the U.S. Army Special Forces. He has served 24 months in Afghanistan and deployed multiple times to South America. He is currently assigned to the U.S. Army John F. Kennedy Special Warfare Center and School as the commander of "Robin Sage", the last phase of the Special Forces Qualification Course. He holds a Master in Military Arts and Sciences in Theater Operations from the U.S. Army School of Advanced Military Studies (SAMS) and an MBA from George Mason University.

Sarah Chana Mocke

Sarah Chana Mocke is a Director in the office of the CTO, Worldwide Services at Microsoft Corporation. She had excelled in many roles at Microsoft in pursuit of broadening her expertise, including those of enterprise architect, solutions architect, sales, program manager and people management. She enjoys the unique insight the sum of her experiences in other roles and disciplines provides and makes use of all of them to foster teamwork and collaboration, to deliver results. She is passionate about connecting, enabling, and empowering others, and greatly prefers working as part of a high performance, cohesive team and wherever possible, contributing to the careers and growth of all of those around her. She enjoys learning and continuously innovating but is biased towards putting ideas into practical actions and results. Sarah has worked with financial institutions, media and entertainment companies, retailers and the public sector to manage delivery of complex projects and in her capacity as an architect to envision, plan, build and deploy large-scale technology implementations. At present, she is responsible for the technical readiness of architecture practitioners across the world within Microsoft Enterprise Services, and for running a large number of communities of practice.

Chris Paparone

Chris Paparone is a retired U.S. Army Colonel who is serving as the Dean, College of Professional and Continuing Education, Army Logistics University, Fort Lee, Virginia. While on active duty, he served in various command and staff positions in the continental United States, Panama, Saudi Arabia, Germany, and Bosnia. He is a proud graduate of the U.S. Naval War College (Command and Staff) and received his PhD in public administration from The Pennsylvania State University, Harrisburg. He and his family reside in the picturesque county-city of Chesterfield, Virginia.

Colleen F. Ponto, EdD

Colleen Ponto is Associate Academic Dean and Faculty of the Organization Systems Renewal (OSR) Graduate Program at the Bainbridge Graduate Institute in Seattle, Washington. Grounded in systems and design, OSR offers a Master of Arts degree in Organizational Leadership. Systems thinking and leadership development is at the heart of Colleen's teaching and consulting and she specializes in helping clients develop their systems thinking skills and leadership capacities in order to address complex organizational issues and opportunities. Colleen also leads a nonprofit organization in her community dedicated to environmental education for children and their families. Before becoming a professor and consultant, Colleen worked as a pulp and paper engineer and production manager for the Weyerhaeuser Paper Company. She received her doctorate in Educational Leadership from Seattle University, her master's degree in Whole Systems Design from Antioch University Seattle, and her bachelor's degree in Paper Science and Engineering from the University of Washington. She lives with her husband and three children near Seattle, Washington.

Skip Rowland

Dr. Skip Rowland is CEO of Banner Cross, a whole systems design learning and development consulting firm, specializing in the challenges of transitioning in an era of rapid and profound change in all our civil institutions. Skip is a Ford Foundation Fellow who focuses on the development and sustaining of an entrepreneurial mindset for business leadership. He is considered a regional expert on economic development, multicultural marketing and cross-cultural and international business policies and practices. Skip has taught business development and entrepreneurship at the University of Washington and Seattle Pacific University and is currently focused on uniting the Puget Sound Region business community with the multiple U.S. Armed Forces/military base leadership and service members to assist the community with absorbing transitioning military service members and helping service members understand the Regional business ecosystem and plan their futures. Skip also works with small to medium sized manufacturing firms to grow their business and enhance their competitive edge.

Skip holds a doctorate in Leadership Education from Seattle University, a master of science in Business Management from Gonzaga University, and a bachelor of science in Social Science from Chapman University.

Ulla Johansson Sköldberg

Professor Ulla Johansson Sköldberg holds the Torsten and Wanja Söderberg Chair in Design Management at Business & Design Lab, a research center at University of Gothenburg. Previously, she was the founding director of the BDL. She received her PhD from Lund University, and held a position as senior lecturer/associate professor at Växjö University. Her interest in art, design and organizations partly comes from her early studies of architecture in the 1970s before working with housing management in 1980s. In the 1990s she became a PhD student and wrote a dissertation about the concept of responsibility in organizations. Her current interest in design and art emanates from her position as professor in design management placed in the faculty of fine and performing art. Since she regards

CPSIA information can be obtained
at www.ICGtesting.com
Printed in the USA
LVOW04*1913260516

490110LV00016B/203/P

9 781623 967451